P9-DBV-389

INTRODUCTION TO ENVIRONMENTAL EPIDEMIOLOGY

Edited by
Evelyn O. Talbott
Gunther F. Craun

Library of Congress Cataloging-in-Publication Data

Talbott, Evelyn.
Introduction to environmental epidemiology / Evelyn O. Talbott, Gunther F. Craun.
p. cm.
Includes bibliographical references and index.
ISBN 0-87371-573-X
1. Environmental health. 2. Environmentally induced diseases—Epidemiology. 3. Environmental monitoring . I. Craun, Gunther F. II. Title.
RA566.T35 1995
615.9′02--dc20 95-2746
CIP

This book contains information obtained from authentic and highly regarded sources. Reprinted material is quoted with permission, and sources are indicated. A wide variety of references are listed. Reasonable efforts have been made to publish reliable data and information, but the author and the publisher cannot assume responsibility for the validity of all materials or for the consequences of their use.

Direct all inquiries to CRC Press, Inc., 2000 Corporate Blvd., N.W., Boca Raton, Florida 33431.

Lewis Publishers is an imprint of CRC Press

International Standard Book Number 0-87371-573-X
Library of Congress Card Number 95-2746
Printed in the United States of America 1 2 3 4 5 6 7 8 9 0
Printed on acid-free paper

Preface

Environmental epidemiology is a fairly distinct area of epidemiologic investigations. While occupational epidemiology is often included, there are many differences that have separated these two study areas. Environmental epidemiology traditionally has dealt with non-occupational exposures and these are often orders of magnitude lower than in occupational environments. Moreover, the populations exposed are larger and more diverse and the effects usually smaller and less easily defined. This has limited opportunities for decisive epidemiologic investigations and greatly increases their complexity.

The importance of environmental epidemiology lies in the large number of people potentially affected and the opportunity for protecting the general population through governmental regulatory activity. Historically, findings from environmental epidemiology have resulted in extraordinary improvements in the health of human populations. The public health movement, which started in the last century, had its roots in studies of the relation between environmental sanitation and health. Perhaps the earliest and what could be one of the most influential epidemiologic studies was the demonstration by John Snow in 1857 that cholera was related to sewage in drinking water. This led to the prevention of cholera and was the kind of information that fueled the movement for improvements in drinking water quality and in general environmental sanitation. Some more recent discoveries from environmental epidemiology are described in the chapters that follow.

The challenge for today's environmental epidemiologists is to identify sensitive populations and sensitive health-related endpoints that can be related to environmental contaminants. Validating and measuring exposures to these contaminants is particularly important and the dictum "no exposure, no disease" cannot be ignored. To date the contribution of environmental epidemiology to standard setting for noninfectious disease has been relatively small. Nearly all environmental standards in effect today are based either on observations in occupational environments, or more often, on extrapolations from animal data. Many question whether this is appropriate. Clearly there is a challenging role for environmental epidemiologists here.

Philip E. Enterline

Introduction

The field of environmental epidemiology can be thought of as a relatively new and burgeoning field. Since the advent of Rachel Carson's, *Silent Spring* in the early 1960's, there has been an increased awareness among both the public and scientific community of the relationship between environmental hazards and human health. With its beginnings in the area of infectious disease, environmental epidemiology has now moved toward quantification of pollutants in air, water and soil through a complex mixture of chemicals, toxins, pesticides and nonbiological agents and the possible associations of environmental exposures with adverse health effects such as cancer, cardiovascular disease, and reproductive outcomes. This book is aimed at those who wish a first look at environmental problems and the epidemiologic approach to the study of the interaction between environmental exposure and disease or other health outcomes.

We have divided the book into major areas of present environmental concern. These include a discussion of the evolution of current American regulatory apparatus with particular emphasis on regulatory and legislative developments within the Environmental Protection Agency. Another discussion area relates to environmental risk assessment — how animal (toxicologic) and human (epidemiologic) data are used in practice to classify risk, the pros and cons of the approach, and the difficulties in quantatively assessing risk.

Causal reasoning in environmental science is important, and the reader is shown by example how associations between exposure and disease should be interpreted and evaluated for possible cause and effect relationships. Both historical and current examples are used with the development of ideas through Henle-Koch and the Bradford Hill Postulates described in detail. Particular problems specifically associated with interpreting results of environmental epidemiologic studies are addressed. Also discussed are the design and analysis of environmental epidemiologic studies with a focus on general principles of statistical reasoning and examples of their application to specific environmental problems including disease clustering and common source epidemics. The mathematics as such will be avoided, and the reader will come away with an increased understanding of why and how statistics are used rather than the ability to use them.

A new technique to the field of environmental epidemiology is biological markers of exposure. An introductory chapter on biological markers of exposure includes what they are, why they are important, and their major uses in environmental epidemiology research. Their relationship with disease state and dose of agent and examples of how rapidly the marker can be measured in the body after the exposure are outlined. Also discussed are criteria to consider when deciding to use a biomarker of exposure in an epidemiological investigation.

The second half of the book discusses epidemiologic aspects, specific examples of environmental exposures and potential risks. Chapter 6 examines some of the issues, epidemiologic methods, and findings that suggest associations between environmental exposures and adverse reproductive outcomes. The author considers, briefly, basic aspects of human reproduction and development and discusses some of the health endpoints that may be associated with exposure to environmental agents, epidemiologic study, approaches to their problems inherent in these studies.

A chapter on risk factors for cancer in the occupational environment is relevant because much of what we know about environmental risks and epidemiologic study design comes from our knowledge of the occupational environment. This discussion concerns the causes of cancer and in particular their association with chemical and physical agents in the occupational environment and is intended to give the reader a brief historical and conceptual view of the problem of environmentally related cancer and how to assess these associations.

The epidemiology of waterborne disease and the importance of drinking water disinfection consists of a brief compendium of infectious waterborne disease risks encompassing both waterborne pathogens and disease, as well as waterborne outbreak statistics in the U.S. and strategies for prevention of waterborne disease. The second portion of this chapter deals with potential health risks of water disinfection and stresses the importance of balancing the microbial by-products and chemical risks of water disinfection.

The chapter dealing with epidemiology of injuries, presents an overview of the public health impact of injury, types and place of injuries, and an epidemiological model developed by infectious disease epidemiology for the study of injury (i.e., agent, the vector or vehicle, as well as environmental factors). The physical hazards of ionizing radiation are reviewed with a discussion of both the known effects of ionizing radiation as well as some of the methodologic issues in environmental radiation epidemiologic studies. The author also includes a brief introduction to nonionizing radiation. Also included in our book is a discussion of areas on specific environmental exposures and health outcomes, a section on electromagnetic radiation, and cancer risks. A critical review of these studies is provided as well as suggestions into better approaches to exposure assessment.

Chapter 12 describes recent progress in the study of childhood lead exposure. Chronicles in the recent progress and understanding of the pathophysiology, epidemiology and long-term consequences of childhood lead exposure in the past decade are presented. A description of lead screening tests and studies of asymptomatic exposure are outlined. The author discusses the behavioral effects of low levels of prenatal exposure to lead and the public health impact of excess lead exposure in our society.

The final chapters (13 and 14) consider health effects from air pollutants and environmental noise exposure. The Clean Air Act requires that sensitive subgroups of exposed populations be protected from adverse health effects of air pollution exposure and the environmental epidemiologic studies discuss the evidence for declines in pulmonary function of children exposed to episodes of high air pollution. The authors of the chapter concerning environmental noise exposure present a short introduction into the major sources of noise, measurement of noise and sound in environment, standards which have been set to control noise, and a description of the most recent epidemiologic studies of several possible health effects of noise from environmental sources.

The editors hope you, the reader, come away with an appreciation of the depth and breath of this newly emerging field of science and an appreciation of its multidisciplinary nature and the difficulty of evaluating effects of environmental exposures on our health status. We would like to acknowledge Dr. Bruce Case of Montreal, Canada who inspired the idea of the possibility of this textbook.

This work was funded in part by a grant from the U.S. E.P.A., number CR812-761.

Evelyn O. Talbott
Gunther F. Craun

Dr. Evelyn O. Talbott is an environmental epidemiologist and associate professor at the University of Pittsburgh, Graduate School of Public Health in Pittsburgh, Pennsylvania. She received her Masters and Doctorate degrees from the Department of Epidemiology from this institution in 1971 and 1976, respectively. She began her career at the University of Illinois in the Department of Preventative and Community Medicine where she helped implement a community-based lead screening program. Her work at the Illinois Institute for Environmental Quality involved researching air quality criteria for lead, cadmium, and mercury exposures. She then returned to the University of Pittsburgh in the fall of 1972 to pursue her doctorate. Her thesis topic involved the study of coronary risk factors and sudden cardiac death in women in Allegheny County, through the Coroners Office. Dr. Talbott continued to work in the area of cardiovascular risk factors but branched out into the area of occupational noise exposure, hearing loss, and blood pressure. She has served on a NIH Consensus Development Conference on Noise and Hearing Loss, completed three major studies of occupational noise exposure, noise-induced hearing loss, and blood pressure in various industrial cohorts, and continues to serve on several national and international committees involved in environmental noise exposure.

Dr. Talbott is the primary instructor for the Environmental Epidemiology Course within the Department and also primary advisor for students in this area. Until recently, she was the Assistant Director of the Center for Environmental Epidemiology, an E.P.A.-supported Center for environmental health effects research. She also served as Secretary-Treasurer of the International Society for Environmental Epidemiology, a 350 member organization dedicated to environmental epidemiology research. She has conducted numerous investigations into the relationship of various environmental exposures and health outcomes. Several of these have involved a population living near a uranium refinery in operation from 1910 to 1937. She successfully traced the individuals who were living within one mile of the plant to determine health outcomes.

Dr. Talbott has published over 45 articles, and she has presented over 50 papers on such topics as the health effects of low-level radiation, noise exposure, and air pollution.

Gunther F. Craun is an environmental epidemiologist and engineer who has been active in water quality and environmental health issues since 1965. He received his education in epidemiology (S.M.) and public health (M.P.H.) at Harvard University. He is a registered professional engineer in the Commonwealth of Virginia and a diplomate in the American Academy of Environmental Engineers. With the U.S. Environmental Protection Agency from its beginning in 1970 until 1991, he directed an epidemiologic research program to assess health effects associated with drinking water contaminants. He founded a consulting firm in 1992 and continues to be active in these issues.

Mr. Craun has served on numerous committees for government, professional and scientific associations, and international organizations to identify potential public health problems, improve water quality, and maintain safe drinking water. He has authored numerous articles in the international scientific, public health, and engineering literature and has edited several books, including *Safety of Water Disinfection: Balancing Chemical and Microbial Risks* (ILSI Press, 1993) and *Waterborne Diseases in the United States* (CRC Press, 1986). The American Water Works Association and the New England Water Works Association have recognized Mr. Craun for his contributions to identifying the causes of waterborne outbreaks and lead and other trace metals in water distribution systems. He received a meritorious service medal from the U.S. Public Health Service in 1992 for his continued work on the surveillance and prevention of waterborne diseases. The E.P.A. also awarded Mr. Craun a meritorious performance citation in 1972 for his participation in the Community Water Supply Study which identified deficiencies in the nation's public water supplies.

Mr. Craun is a member of the American Water Works Association, American Public Health Association, Society for Epidemiologic Research, and International Society for Environmental Epidemiology. He also serves on the Editorial Advisory Board for the *Journal of the American Water Works Association* and the Research Advisory Council, American Water Works Association Research Foundation.

Contributors

Bert Brunekreef
Department of Environmental and Tropical Health
Agricultural University
Wageningen, The Netherlands

Gunther F. Craun
Global Consulting for Environmental Health
Radford, Virginia

Douglas Dockery
Harvard School of Public Health
Boston, Massachusetts

Benjamin G. Ferris, Jr.
Harvard School of Public Health
Boston, Massachusetts

Irva Hertz-Picciotto
Department of Epidemiology
University of North Carolina
Chapel Hill, North Carolina

Barbara S. Hulka
Department of Epidemiology
University of North Carolina
Chapel Hill, North Carolina

Erick K. Ishii
Department of Epidemiology
Graduate School of Public Health
University of Pittsburgh
Pittsburgh, Pennsylvania

Patrick L. Kinney
New York University Medical Center
Tuxedo, New York

William V. Luneburg
School of Law
University of Pittsburgh
Pittsburgh, Pennsylvania

Gary M. Marsh
Department of Biostatistics
University of Pittsburgh
Pittsburgh, Pennsylvania

Herbert L. Needleman
University of Pittsburgh
Pittsburgh, Pennsylvania

Leon S. Robertson
Nanlee Research
Branford, Connecticut

Lowell E. Sever
Battelle Pacific Northwest Laboratories
Battelle Seattle Research Center
Seattle, Washington

Jack Siemiatycki
Armand Frappier Institute
University of Quebec
Laval, Quebec, Canada

Frank E. Speizer
Channing Laboratory
Department of Medicine
Harvard School
Bringham and Woment's Hospital
Boston, Massachusetts

John D. Spengler
Harvard School of Public Health
Boston, Massachusetts

Daniel J. Strom
Health Protection Department
Battelle Pacific Northwest Laboratories
Richland, Washington

Evelyn O. Talbott
Department of Epidemiology
Graduate School of Public Health
University of Pittsburgh
Pittsburgh, Pennsylvania

Gilles Thériault
Department of Occupational Health
Faculty of Medicine
McGill University
Montreal, Quebec, Canada

Shirley Jean Thompson
Department of Epidemiology and
Biostatistics
School of Public Health
University of South Carolina
Columbia, South Carolina

Neal D. Traven
Department of Epidemiology
Graduate School of Public Health
University of Pittsburgh
Pittsburgh, Pennsylvania

Marilyn F. Vine
Department of Epidemiology
University of North Carolina
Chapel Hill, North Carolina

James H. Ware
Harvard School of Public Health
Boston, Massachusetts

Table of Contents

CHAPTER 1

The Legal Context of Environmental Protection in the United States

William V. Luneburg

I. HISTORICAL BACKGROUND AND OVERVIEW

The movement for increased protection of the environment, which gained particular prominence in the late 1960s in the U.S., and which resulted in the exponential growth of federal and state regulatory structures, represented the confluence of two separate social forces, each with its own long and distinguished history.[1] "One of these was essentially conservationist, outdoors oriented, intent on protecting our scenic resources and natural heritage for ourselves and future generations. The concern for 'quality of life' ... has been most strongly associated with that part ...."[2] John Muir, founder of the Sierra Club, is one of the most prominent persons associated with this interest. The other part of the movement drew its support from persons whose primary concern was "in public health, in protecting people in our cities and highly populated areas against the pervasive impact of environmental pollution" of various types.[2a] The discussion that follows in this section focuses largely on this second stream of policy development.

With the growth of industrialism and the mushrooming of urban population centers after the middle of the 19th century, water supply and sewage disposal problems became acute. Moreover, coal-burning furnaces fouled the air with soot and cinders. Public fear of contagion ultimately became a major influence on the formation of public policy.[3] Beginning in 1857, doctors on the east coast started to meet every year to develop recommendations for improved sanitary conditions.[4] At these meetings the creation of state and local boards of health was favored as a means of regulating the establishment of quarantines, collecting vital statistics, and improving public sanitation.[5] With the general acceptance that waterborne germs (rather than sin) caused various diseases, provision of clean drinking water became the first priority for public health officials. Since sewage could be treated to kill bacteria and to remove organic contaminants, provision of adequate sewage treatment also be-

0-87371-573-X/95/$0.00+$.50

came a focus of attention, though for economic reasons this strategy was only slowly adopted.[6]

With regard to drinking water, the historical linkage between activity in the public health sphere and what is considered modern environmental legislation — in this case the Safe Drinking Water Act of 1974 — is clearly shown by the development from what are now entitled the *Standard Methods for the Examination of Water and Waste Water*, to the Public Health Drinking Water Standards and, finally, to the United States Environmental Protection Agency's drinking water regulations.[7] Federal involvement with the problem of safe drinking water began in 1878 with the enactment of the first National Quarantine Act (to prevent the introduction of infectious or contagious diseases into the U.S.).

After the beginning of the 20th century, public health officials increasingly pointed an accusing finger at industrial water pollution, but that soon became viewed less of a "health" problem than one of "conservation".[8] Accordingly, a major portion of the authority for controlling industrial discharges into waterways was initially vested in departments of conservation and natural resources.[9] In fact one of the first federal ventures into water pollution control, the Refuse Act of 1899, was designed to protect navigability of waterways. While the federal Oil Pollution Act of 1924, which prohibited willful and grossly negligent discharges of oil, was concerned with health effects as a subsidiary purpose, it too was primarily motivated by the commercial effects of pollution.[10] By 1948, however, this situation began to change as Congress adopted the Water Pollution Control Act, the purpose of which was primarily to provide federal funds for municipal sewage treatment, but also authorized federal mediation and enforcement in cases where interstate water pollution threatened public health.[11]

With regard to air pollution, the first legislative bodies to take affirmative action in the later part of the 19th and early part of the 20th centuries were the common councils of large industrial cities. If states acted at all, it was by authorizing cities to regulate smoke nuisances.[12] Local ordinances took a variety of forms. Some simply declared that the emission of dense smoke from any chimney or smokestack was a public nuisance and imposed a modest fine on this activity. Others placed an affirmative duty on the owner of the facility emitting the smoke to eliminate the emissions. Finally, since the extensive use of soft, high-sulfur, bituminous coal was the root cause of the air pollution problem in many areas, some ordinances flatly prohibited the importation, sale, and use of coal of certain ash and sulfur content.[13] Most of these ordinances, however, lacked provisions for effective enforcement, though, if an administrative agency was employed, it was often the board of health or the building inspection department.[14] If abatement was to occur, members of the public who were injured were forced themselves to resort to the courts under nuisance theories, which sometimes permitted pollution to continue because of the large economic costs of effective abatement.[15]

In view of the fact that air pollution problems only increased during the first half of the 20th century, lawmakers tried new approaches. County air pollution agencies replaced city smoke inspectors.[16] Moreover, the federal government became involved with the Air Pollution Act of 1955, which was designed primarily to promote research into the area of air pollution. The Clean Air Act of 1963 finally provided the Department of Health, Education and Welfare with some limited jurisdiction to abate interstate air pollution problems[17] (See for a detailed history of air pollution control in the United States).

In sum, prior to the 1960s, environmental protection viewed as a matter of public health was, with some exceptions, left largely to the states. Generally speaking, the federal government's role was limited to grant-making and research assistance with some jurisdiction to deal with interstate pollution problems. Courts adjudicated private law cases of air, water, and other environmental contamination under nuisance, trespass, and other theories and rendered judgments that were limited to the parties before the court. State and local governments created regulatory regimes — including permit systems applicable to new and/or existing sources — though their effectiveness in eliminating pollution was rarely demonstrated in practice.[18]

However, starting in 1970, in response to increased public alarm over the condition of the national and global environments, Congress enacted over the next 20 years an increasing number of statutes which, in both their detail and technical complexity, challenge the comprehension of even the best technically trained professionals and the administrative skills of the most accomplished bureaucrats. Statutory requirements are elaborated in thousands of pages of regulations, guidelines, staff manuals, and other agency-generated documents. While some discretion remains in the states, the federal government sets the general and, in many instances, the specific requirements applicable to the various administrative programs for pollution prevention and cleanup. States can usually establish more stringent standards. But they are bound by the federal standards as a "floor" below which they may not fall.

The system — or, more accurately, the systems — for environmental protection have not been statutorily or, as yet, administratively integrated to deal effectively with the intermedia characteristic of environmental contamination. One method of pollution control often shifts hazardous materials from one area (e.g., the air) to another (e.g., the surface water). In other words, the specific focus is not on aggregate risk reduction in a comprehensive and coordinated fashion.[19] If a discharge of pollution is to the ambient air, the Clean Air Act controls; to the surface waters, the Clean Water Act applies; to the groundwater, the Resource Conservation and Recovery and the Safe Drinking Water Acts are the touchstones for action. While some commonalities in approach unite many of these statutory frameworks, the detail and peculiar features of statutory and administrative regulation (along with bureaucratic turf-fighting) often foreclose efforts in the direction of integrated treatment.

Increasingly, attention is turning to pollution prevention; that is, efforts to avoid the use of products or processes that may degrade the environment.[20] This is because of the very substantial costs of cleaning up pollution damage after it happens and the inability to achieve a full remediation in many instances. Federal statutes dealing with pesticides and toxic substances, though ineffective to a degree and for a variety of reasons, exemplify this approach.

The sections that follow set forth in some detail: (1) the background against which the federal and state presence has grown in the area of environmental protection, (2) a general typology of the regulatory and other systems that have been established, and (3) an overview of the specific provisions of some of the most important federal statutes that directly concern themselves with environmental control.

II. CONSTITUTIONAL BACKGROUND

The constitutional division of governing authority in the U.S. between the central (federal) and the state governments profoundly affects the nature and scope of environmental regulation. States retain all those powers not expressly granted to the federal government or necessary and proper to carry out the powers of that government. The authority of Congress to "provide for the ... general Welfare of the United States" and "regulate Commerce with foreign Nations, and among the several States ..."[21] are the grants of power in Article I of the United States Constitution on which most modern environmental regulation is founded.

The first of these authorizes, for example, grants-in-aid to the states conditioned on their compliance with congressionally determined goals and procedures (for example, grants for the construction of sewage treatment facilities). In other words, states can be offered financial incentives, which, however, can be withdrawn where there is substantial noncompliance with grant conditions. Such conditions must, however, bear some rational relationship to the purpose of the federal spending program.[22]

The second font of power is the basis for statutory programs that directly command and regulate the activities of state government and private parties. Since the 1930s, the Commerce Clause has been construed so broadly that today Congress can legitimately assert jurisdiction over any activity that has or may have a substantial effect on interstate and foreign commerce (and what activities do not today?).[23]

Where Congress or a federal agency has acted within the scope of their granted powers, its legislation and regulations displace (under the Supremacy Clause)[24] inconsistent state law along with state law that substantially interferes with the implementation of federal policy. In many instances, the scope of this "preemption" is far from clear. The presumption is generally against preemption because of the historic roles of the states as protectors of public health and welfare and the legitimate local interests at stake in regulation. Conventionally, the problem is treated as one of ascertaining "congressional intent", something that only occasionally is easily discovered. In fact, in recent years many federal environmental statutes, such as those dealing with nuclear plant licensing[25] and oil pollution prevention,[26] have presented courts with difficult tasks in ascertaining the degree to which state regulatory authority has been preserved. The establishment of tailpipe emission standards for motor vehicles was, for a long time, an area entirely reserved for the federal government (except in the case of California, whose ozone problem seemed to suggest more state flexibility). The "preemption defense" is often invoked today in litigation by private businesses in their attempts to escape state or local laws that they find unduly burdensome.

Even where Congress has not enacted legislation, state attempts to regulate an area may be foreclosed by "Dormant Commerce Clause" doctrine. According to the Supreme Court, one of the purposes of the framers of the federal Constitution was to avoid "economic isolation" and "protectionism" by the previously fully sovereign states.[27] Accordingly, state or local statutes and regulations that may burden interstate commerce are subjected to careful judicial scrutiny to determine whether that impact is justifiable. Recently, attempts by states to limit the importation of solid and hazardous wastes have provided occasion for much of the important litigation in this area. In dealing with these cases the Supreme Court has laid down three rules.

First, state regulation that directly regulates the importation from other states of waste by forbidding that importation or imposing different disposal fees with regard to in-state and out-of-state waste is generally invalid per se.[28] Secondly, however, if the out-of-state waste presents different types of hazards or requirements for disposal than waste generated within the state, the receiving state may impose an outright ban on importation or otherwise regulate the disposal differently than in-state generated waste. But the receiving state carries the burden of establishing how that differential regulation is clearly rooted in the differing characteristics of the in-state and out-of-state generated waste.[29] Finally, if a state or local government regulates, without overt discrimination, the disposal of in-state and out-of-state waste but the regulation has, nevertheless, a substantial impact on interstate commerce, the regulation will be upheld to the extent there is a legitimate local health or other public interest served by the statute or ordinance and the effects on interstate commerce are not the purpose of the regulation.[30]

The Dormant Commerce Clause doctrine recognizes that Congress can, if it wishes, permit a state to enact legislation barring or otherwise restricting the importation of out-of-state materials. It has done so in a variety of statutes, including the Low-Level Radioactive Waste Policy Amendments Act of 1985.[31]

In structuring federal environmental programs, Congress has developed an approach called "cooperative federalism", which reflects many of the foregoing constitutional principles. In recognition of the important role of the states in the federal union, their legitimate local interests, and the traditional involvement of state and local government in protecting public health and welfare, Congress may offer the states the option of designing measures for environmental protection — subject to specific federal restrictions and oversight. State regulation may be more, but not less, stringent than the baseline established by Congress (or its administrative delegate, usually the United States Environmental Protection Agency [EPA]). If the states fail to design federally approvable programs, Congress authorizes a federal agency to adopt and enforce a federally imposed program (implementation of which may be, in appropriate cases, delegated to the states). Inconsistent state regulation is preempted. This is the pattern, for example, of the Clean Air Act and the Clean Water Act.

In exercising its expansive authorities, Congress is, however, subject to various constitutional restrictions, only some of which are typically relevant to environmental regulation. First, while Congress may offer financial incentives to states to encourage them to regulate for the purpose of achieving congressional policy goals or, alternatively, preempt state law that interferes with achievement of those goals, it may not directly command states to enact and implement regulatory schemes.[32] Second, while the scope of the Commerce Power is indeed broad, congressional (and state) regulation of private party activity is subject to the Takings Clause, which requires the government to compensate private parties whose property is "taken" for public purposes.[33] Sometimes such regulation eliminates so much of the value of private property that, according to the Supreme Court, compensation is due. Over the last decade, "takings" challenges to environmental restrictions have increased. The Supreme Court's decisions have, in fact, encouraged this phenomenon. Wetlands protection has been an area of particular activity, as where the Corps of Engineers denies a permit to fill an area found to be a "wetland". The restrictions thus imposed on development can allegedly deny "all economically beneficial or productive use of land", the constitutional test for a taking.[34]

One of the persistent complaints about government intervention for environmental protection is that administrators fail to consider the social costs of regulation and therefore command results that are not "economically efficient". The Takings Clause of the Constitution requires, at least in some instances, that the government absorb some of the costs of its regulatory efforts. The hope for some is that this will act as a prophylactic against over-regulation.

III. THE COMMON LAW SYSTEM OF REGULATION

Even without statutes to apply and enforce, state courts have traditionally sought to regulate environmental hazards of various kinds by issuing injunctions to halt private activity or entering judgments for damages. The theories employed have generally fallen into four categories: (1) no one may use his or her property to substantially interfere with the use of another's property or various public rights (private and public nuisance); (2) no one may act as a "reasonable person" would not act and, in the process, cause injury to another (negligence); (3) no one may directly interfere with another's exclusive possession of his or her property without permission from the landowner (trespass); and, finally, (4) no one may engage in high risk-creating activities in areas where they are not generally or appropriately found (strict liability).

Courts have applied these theories in cases of air, surface water, and groundwater pollution. However, traditionally this case-by-case approach has been found wanting for a variety of reasons. Some of these reasons are structural in nature: outside the context of so-called class actions (where many injured people join as plaintiffs in one lawsuit), courts lack the ability and, allegedly, the legitimacy to create remedies that adequately address systemic problems having numerous sources and possible solutions; courts generally intervene only after damage is done; and, finally, courts lack substantial technical expertise that may be necessary for formulating a specific or more general remedy.

Other limitations to the common law method relate to the elements of the theories of action that the courts have evolved over time. Most importantly, where there may be many sources for a particular problem of environmental contamination, the rules of causation and proof may disfavor recovery by injured parties. Even where there is only one possible source of the harm, those same rules may be insurmountable or the chain of causation deemed to be too attenuated as a matter of policy judgment by the court. Moreover, the types of environmental damage for which there can be compensation under the common law may be limited. General ecosystem harms may, for example, escape remediation and awards of compensation.

In recent years the courts have tried to accommodate their traditional approaches to the peculiar problems of proof and damage presented by so-called toxic torts. Burdens of proof linking alleged health effects to particular causes have been relaxed to a degree, in recognition that the etiology of disease may be much disputed by reputable scientists or relevant scientific study may be incomplete. Moreover, the courts have been willing to cope with the long latency period of some environmentally caused diseases by ordering medical surveillance in cases where a plaintiff or a group of plaintiffs have been exposed to a hazardous substance. Finally, under the so-called "public trust" doctrine, governments have been given the right (and duty in some instances) to protect public resources and ecosystem integrity.

IV. THE LEGISLATIVE AND ADMINISTRATIVE RESPONSE TO ENVIRONMENTAL CONTAMINATION

A. HISTORICAL RETROSPECTIVE

Because of the institutional limitations of common law litigation, perceived deficiencies in judicial doctrines, and also for political reasons, since the last third of the 19th century legislatures have turned increasingly to the administrative bureaucracy to fashion and implement solutions to pressing social ills. The Rooseveltian New Deal represented a culmination of sorts in this development, with the establishment of numerous agencies to regulate the manifold aspects of the national economy and to distribute benefits to needy individuals. There was in the process a clear shift in the balance of power between the federal government and the states as public policy issues were conceived in "national" rather than local terms.

Two viewpoints came to exist in an uneasy balance during this period: on the one hand, the notion of the agency as the embodiment of "technical expertise" and, therefore, "above politics", and, on the other, a concern for the protection of "law" and "legal rights", particularly the rights of private enterprise. To accommodate these potentially divergent notions, the statutory and common law doctrines of administrative law developed rapidly during the 1930s and early 1940s, culminating in the enactment in 1946 of the federal Administrative Procedure Act.[35] Moreover, that "law" has continued to evolve as, during different historical periods, one of these viewpoints has dominated the other.[36]

Table 1 Significant Federal Environmental Protection Statutes since 1969[a]

1969	National Environmental Policy Act
1970	Clean Air Act Amendments
	Reorganization Plan No. 3 (1970) (U.S. Environmental Protection Agency created)
1972	Federal Environmental Pesticide Control Act (amendments to the Federal Insecticide, Fungicide, and Rodenticide Act)
	Federal Water Pollution Control Act Amendments
1973	Endangered Species Act
1974	Safe Drinking Water Act
1976	Resource Conservation and Recovery Act
1977	Clean Air Act Amendments of 1977
	Toxic Substances Control Act
1980	Comprehensive Environmental Response, Compensation and Liability Act
1984	Resource Conservation and Recovery Act Amendments
1986	Emergency Planning and Community Right-to-Know Act
	Superfund Amendment and Reauthorization Act
1987	Clean Water Act Amendments of 1987
	Water Quality Act
1990	Clean Air Act Amendments
	Oil Pollution Act
	Pollution Prevention Act

[a] The list includes some, but not all, amendatory legislation.

Starting in 1970 another floodtide of legislation began as new environmental and public health statutes were enacted at both the federal and state levels, with the central government generally taking the lead and setting the parameters for permissible state action (see Table 1). While some of these statutes directly enlarged the regulatory authority of the courts, most were structured in such a way as to place primary policy-making and implementation powers in administrative agencies. The courts were utilized largely as forums where deferential review of agency policy-making and adjudication took place, as well as where civil and criminal enforcement of environmental regulations occurred.

Congressional concern for the expeditious protection of public health translated into very specific statutory time deadlines for agency action in adopting regulations and achieving environmental goals. The Clean Air Act was typical, requiring that the national ambient air quality standards for ozone, sulfur oxides, and other widely distributed pollutants be attained by the mid-1970s.

Moreover, for the 24-year period following the inauguration of President Nixon in 1969, the policy-making institutions of the federal government (the administrative bureaucracy and the Congress) were dominated by different political parties for all but 4 years (the Carter administration). Divided government was characterized by disagreement, suspicion, and distrust, which were fueled in part by EPA's inability or unwillingness to conform to exacting legislative mandates for environmental protection. Accordingly, Congress took it upon itself to draft increasingly specific legislative provisions to reduce or, in many cases, eliminate administrative discretion.[37] This trend reached its apex in 1990 with the enactment of amendments to the Clean Air Act, which amendments consisted of several hundred pages of detailed specifications and directions with regard to federal administrative and state action.

Also, during this same period, Congress routinely included in environmental statutes so-called "citizen suit" provisions that authorize private persons to sue EPA to force it to comply with congressional directives.[38] Citizens are also empowered to bring actions for injunctive relief and civil penalties against public and private entities that violate federally approved or imposed environmental regulations. In this way, underfunded agency enforcement efforts can be supplemented and recalcitrant federal and state agencies either forced to act or their inaction displaced by citizen initiative.

Distrust of administrative agencies as unduly subject to the influence of regulated parties, along with the recognition that citizens have a bona fide interest in the protection of various environmental amenities, both contributed to the development of various judicial doctrines of administrative law after 1970. For instance, courts required agencies to permit increased citizen intervention in administrative proceedings; relaxed requirements that had previously barred citizens from seeking judicial review of administrative action (e.g., standing); determined that failure to regulate presented a justiciable issue; and required agencies to take a "hard look" at the need for various environmentally protective measures, while at the same time judicially scrutinizing agency decisions with some care for arbitrariness.[39]

During the 1970s and 1980s congressional and agency sophistication with regard to environmental problems and regulatory approaches grew. The national economy also experienced periods of low growth and high unemployment. As a result, Congress, EPA, and the White House stressed the need for benefit/cost analysis in crafting environmental rules, in order to avoid the imposition of unnecessary or disproportionate costs on the economy. Sometimes mandates to consider economic costs were written directly into legislation. On the executive side, this trend was formalized by the Reagan Administration's adoption of an executive order (12,291) mandating that each federal agency conduct a benefit/cost analysis for each major rule proposed and that the Office of Management and Budget (OMB) review and comment on these analyses and the subject rules.

The unquantified or unquantifiable nature of many benefits of environmental regulation (for example, the value of *clear* air in national parks) made the application of benefit/cost analysis problematic. Moreover, the professed deregulatory bias of Republican administrations created suspicions on the part of Congress, the agencies, and environmentalists that the administration of Executive Order 12,291 was manipulated to achieve a political agenda that both undervalued environmental protection and disregarded statutory requirements. As a counterpoint, for its part, OMB charged that risk assessment and management techniques — which were often expressly required by Congress in environmental statutes — were allegedly employed by agencies in such a way as to build overly protective "margins of safety" into environmental regulation.[40] It was almost as if "worst case" assumptions with regard to health and environmental effects were used as a rather crude counterweight to the difficulties or impossibility of quantifying the nature and weight of many environmental benefits in public decision making.

B. ADMINISTRATIVE LAW AND PROCEDURE

Agency functions are generally categorized as either "rule making" or "adjudication" for the purpose of determining the types of procedures that apply to agency decision making. The source for those procedures include the Due Process Clauses of the Fifth and Fourteenth Amendments to the United States Constitution, specific statutes establishing agency programs, administrative procedure acts, and agency regulations.

Rule making generally connotes policy making for a general class of regulated entity. Only minimal procedures attach, including (at the federal level) notice of a proposed rule published in the *Federal Register*, solicitation of public comment, and issuance of the final rule with an explanatory statement.[41] Adjudication involves the application of a statutory or administrative rule to a particular set of facts. Generally speaking, more elaborate procedural protections apply to this process, including the right to present evidence and to cross-examine witnesses.[42]

Many agencies have both the power to make rules with the force of law and to conduct adjudications, with the role of the courts to assure that the ultimate agency decision is consistent with constitutional and statutory norms and avoids arbitrariness and abuses of discretion. Judicial deference to agency determinations depends, among other things, on the type of issue (law, fact, or policy) resolved by the agency.

Under some statutes, an agency is under an obligation to make policy by rule. Particularly in the environmental area, there are likely to be congressional deadlines for this required rule making. In other instances, agencies can generally create policy by rules or in the course of individual adjudications, or both.[43]

Increasingly, statutory schemes for environmental protection have adopted the administrative technique of requiring permits for new and existing sources of pollution and prohibiting releases to the ambient environment without a permit or in violation of the permit's terms or conditions. Adopted by the Federal Water Pollution Control Act Amendments of 1972, this regulatory technique is employed in the Resource Conservation and Recovery Act with regard to the disposal of hazardous waste and, most recently, in the 1990 amendments to the Clean Air Act. A permit system can serve a variety of functions, including the identification of the universe of those facilities subject to regulation and taking inventory of their discharges for environmental quality planning purposes. A permit also eliminates uncertainty for the pollution source as well as enforcement authorities by collecting in one place (the permit) all applicable environmental requirements. Typically, other than discharge limits, permits include compliance schedules, reporting and monitoring requirements, and provisions dealing with malfunctions. Monitoring reports prepared by sources under the terms of their permits are required to be publicly available and thus often form the basis for citizen suits.[44]

In environmental statutes, which are characterized by the "cooperative federalism" approach, Congress gives the states the option to establish and administer permit systems, subject to federal approval of their specific features. The issuance and amendment of permits constitutes agency adjudication and thus must be accompanied by rather elaborate pre-decisional hearings in which citizens and affected states may participate. Review of agency issuance or denial of a permit may occur both at the state and federal administrative and judicial levels, thus promising lengthy and costly delays in controversial cases.

In conclusion, it should be noted that, procedural niceties aside, federal and state agency policy making do not occur in a political vacuum. The chief executive's appointment and removal power as it applies to high agency officials, rule-review mechanisms such as Executive Order 12,291, legislative oversight, and appropriations hearings, among other things, all operate to bring nontechnical factors into play at the agency decision-making stage.[45] (See Strauss for an excellent overview of the administrative law system in the U.S.)

V. ENVIRONMENTAL DECISION MAKING IN THE FACE OF PERVASIVE UNCERTAINTY

The goal of the scientific method is to arrive at accurate conclusions. Likewise, the legal system aspires to accuracy in at least one of its functions, that of adjudicating the facts in individual cases (e.g., did he run the stop light?). However, to an important degree, the similarity ends there. The legal system demands a "winner" and a "loser" in adjudicated cases, even in the face of disputed facts. Factual uncertainties are depositively resolved (for the purpose of the litigated case) one way or the other, often with the assistance of presumptions and burdens of proof. Though they may be employed in evaluating the relevance and reliability of trial evidence, confidence intervals, tests for statistical significance, and the concept of mathematical probability are often considered by the courts as secondary to the need for dispositive adjudicatory conclusions.[46] In short, the law must at some point take that "leap of faith" from available evidence to final decision.

When it comes to determining what the administrative "rules" in the environmental area should be (as in the case of establishing various ambient air or other environmental standards), both legislatures and the courts concede that decision making must operate in the face of pervasive uncertainty ("on the frontiers of science"), which is qualitatively different from the type of factual uncertainty characteristic of the usual tort and contract case. Often there must be a legal decision — whether and how to regulate. It is openly conceded that judgments based on scientific studies whose conclusions are either disputed or qualified are more appropriately considered to be in the realm of policy making, not "fact finding". Accordingly, scientific expertise is important — but only up to a point. Moreover, Congress often establishes precautionary presumptions in favor of avoiding adverse health and environmental effects and available scientific evidence must be evaluated in light of those.

It is this "nonfactual" aspect of environmental decision making that casts in high relief the traditional tension in administrative law between, on the one hand, a vision of agencies as "experts" above "politics" who should be left alone to carry out their missions and, on the other hand, a concern for the protection of "law" and "legal rights" of the regulated parties. More specifically, to the extent that environmental rules represent at least as much, if not more, guesswork and policy making as "fact-finding", it is argued that unelected bureaucrats should be subject to the supervision and control of some elected official (e.g., the chief executive).[47] Since the courts lack a direct electoral mandate, they may not be considered appropriate "guardians" in this context. At the federal level, rule making review by the OMB or other entities close to the President has been employed to supply this form of executive control, though in some quarters the review process itself has been criticized for subverting various procedural and substantive legal constraints on agency action.

VI. TYPOLOGY OF ENVIRONMENTAL STATUTES

Legislatures have been most creative in fashioning strategies to cope with environmental contamination. It is possible, however, to catalog here those strategies that most often find expression in existing or proposed laws. It should be noted that several important statutes, such as the Clean Air Act, embody a combination of many of these approaches, which are often structured to reinforce one another.

Provision of information to the public regarding health and environmental effects — Various federal statutes, such as the Federal Insecticide, Fungicide, and Rodenticide Act,[48] require that certain products bear warnings to the ultimate user with regard to the health and other hazards presented by the product. The Emergency Planning and Community Right-to-Know Act[49] requires, among other things, that companies file toxic chemical release forms and emissions inventories with EPA and the states each year, which must include estimates of the annual quantity of toxic chemicals entering each environmental medium. This information is available to the public.

Taxation of disfavored activities and other charges imposed on polluting activities — Economists have favored the use of emission charges to discourage pollution by forcing manufacturers to absorb the true costs to the environment of their activities and, accordingly, to seek least-cost control techniques. While not a frequently invoked technique in the United States, Congress has employed it on occasion, as it did in 1989 by imposing an escalating tax on the production of chlorofluorocarbons to accelerate the phaseout of these ozone-depleting chemicals and to encourage the development of substitutes. In 1990, with the enactment of the amendments to the Clean Air Act, which require the creation of state permit systems for new and existing air pollution sources, Congress mandated that permittees pay at least $25 per ton of emissions to the permitting agency. These fees will be used to defray the costs of administering these permit programs.[50]

Subsidies — In the past the government has both directly and indirectly subsidized various environmentally questionable activities. For example, timber harvesting and grazing on public lands have been permitted for years without charging the true market costs for these benefits. Conversely, the federal government has distributed funds to be used to improve the environment, as in the case of grants to cities to build and improve mass transit systems like bus lines and subways which may reduce automobile pollution.

Impact assessment — The first, and in many ways, one of the most important of the modern environmental statutes is the National Environmental Policy Act (NEPA) of 1969. It requires that, with regard to major federal actions that may have a significant impact on the environment, all federal agencies prepare and consider in their decision making an analysis that canvasses the environmental impacts of federal actions, along with the alternative courses of action that may mitigate those impacts. NEPA has been copied extensively by the states and in other countries. It basically represents a "stop and think" approach to environmental protection.

Regulation of market access — It is often cheaper to prevent pollution than to clean it up once it has occurred. Both the Federal Insecticide, Fungicide, and Rodenticide Act[51] and the Toxic Substances Control Act[52] authorize premarket review of new products that may present an unreasonable risk to human health or the environment and, if necessary, restrictions on the production and use of such products. The Delaney Clause of the Federal Food, Drug, and Cosmetic Act[53] prohibits food and other additives that have been found to induce cancer when ingested by man or animals — with no *de minimis* levels permitted.[54]

Liability and insurance schemes — At the other extreme are those statutes that impose liability for environmental damages that have already been caused. Perhaps the most famous (or infamous, depending on your point of view) is the Comprehensive Environmental Response, Compensation and Liability Act,[55] which imposes the costs for cleaning up sites used for the disposal of hazardous substances on present and former owners of these disposal sites, along with generators and transporters of the substances.

Technology/performance standards — A statute or agency regulation may impose limitations on discharges or emissions to the ambient environment, which limits are drawn from those attained by certain technological processes for pollution control or reduction. With regard to automobile-created pollution, Congress specified permitted amounts of tailpipe emissions of carbon monoxide, hydrocarbons, and nitrogen oxides and basically directed the automobile industry to develop the necessary technology to meet them ("technology forcing"). More commonly, the technology on which discharge limits is based is "available" in the sense that it is off the drawing board and employed somewhere, though perhaps to a very limited degree. Technology-derived standards vary in the degree to which they are based on more commonly employed processes and controls, as well as with regard to the relevance of economic cost. These standards may be imposed on new and/or existing sources, with those applicable to new sources being generally more stringent than those applied to sources that will have to retrofit facilities. A welter of acronyms is used to refer shorthandedly to these various standards. In the context of the Clean Air Act, for example, these include MACT (maximum achievable control technology), LAER (lowest achievable emission rate), BACT (best available control technology), GACT (generally available control technology), and RACT (reasonably available control technology).[18]

Environmental quality standards — Water quality and ambient air quality standards describe with more or less specificity the concentrations of pollutants permitted in the ambient receiving body that are deemed protective of public health, welfare, and biosystems generally. Such regulations are usually not directly enforceable. Rather they are implemented by the imposition of specific emission or discharge limits on individual sources of pollution. In the acid rain program created in 1990, where the environmental goal to be attained is defined directly in terms of the tonnage of sulfur dioxide permitted to be emitted nationally each year, pollution allowances (up to this amount) issued to existing emission sources can be bought and sold on the open market. Theoretically, in that way facilities that can cheaply reduce their emissions more than legally required can sell the "surplus" emission allowances to others that find it more expensive to reduce pollution to permitted levels. Thus, the environmental goal will allegedly be achieved in the most cost-effective manner, with overcontrol eliminated. These same concepts are now being, or may be applied in the future, with regard to programs to attain the national ambient air quality standards,[19] to achieve required water quality levels, to preserve wetlands, and to reduce the need for hazardous waste disposal facilities, among other environmental goals.

VI. AN OVERVIEW OF SOME OF THE PRINCIPAL FEDERAL ENVIRONMENTAL STATUTES

Federal and state environmental statutes and implementing regulations fill thousands of small-print pages. What is attempted here is but a summary of the general outlines of what are the more commonly encountered federal schemes for environmental regulation.

A. THE NATIONAL ENVIRONMENTAL POLICY ACT[58]

All federal agencies are directed to consider the environmental effects of their activities, including the policies that they adopt and attempt to implement. At the time of a proposal for a "major" federal action that may have a significant impact on the environment, the agency must have prepared a final environmental impact statement (EIS) that canvasses the beneficial and adverse effects, the alternatives to the action proposed, and the ways to mitigate identified adverse effects. Prior to the final EIS stage, agencies prepare environmental assessments (EAs) to determine whether the action to be proposed is one requiring a full-dress EIS. If an agency determines that its proposal does not pass the threshold for EIS preparation, it issues what is called a FONSI (a finding of no significant environmental effect). A draft EIS is circulated among concerned federal agencies for their comments prior to issuance of the final EIS, which by law must accompany the proposal through the agency decision making process. The purposes of the NEPA impact statement are twofold: (1) to help assure that agencies in fact consider the environmental effects of their actions and (2) to provide the President, Congress, and the public generally information that they may need to evaluate and respond to proposed agency actions.

B. THE ENDANGERED SPECIES ACT[59]

The Endangered Species Act requires the identification and listing of endangered and threatened species of plants and animals, along with the habitats necessary for their survival. Following the listing, no federal agency may take any action that jeopardizes the continued existence of the species. Consultation with the Fish and Wildlife Service in the Department of the Interior is required as part of the effort to avoid that effect. There is, however, a provision exempting federal agencies from this prohibition (the so-called "God" Committee process), though this exemption is subject to stringent procedural and substantive restrictions. Moreover, no person or agency can "take" (e.g., injure, harass, or kill) a member of a listed species, subject to very narrow exceptions, including what are called "incidental take" permits, which, however, may not be issued if they will reduce the likelihood of the survival of a listed species.

C. THE FEDERAL INSECTICIDE, FUNGICIDE, AND RODENTICIDE ACT[60]

This statute requires registration of so-called economic poisons (pesticides) with EPA prior to their marketing. In order to obtain registration, the applicant must submit to EPA data and other information relating to the safety and efficacy of the product so that EPA can determine the proper contents of the labeling, the likelihood of unreasonable adverse effects on the environment, and the need for restrictions on the use of the product. Following registration, the use, sale, and distribution of a pesticide may be halted on an emergency basis or after expedited suspension or more lengthy registration cancellation proceedings where the EPA determines that the continued usage of the product poses or may pose a threat to human health or the environment.

D. THE TOXIC SUBSTANCES CONTROL ACT[61]

With regard to chemicals currently on the market, EPA may require testing for their potential for "unreasonable risk of injury to health or the environment" and, if such a risk is found, the agency may limit or forbid the manufacture, processing, and distribution of the product, require changes in labeling, or regulate the manner of disposal of the substance.

With regard to new products coming on the market, manufacturers must give EPA advance notice of their manufacture and sale and, in appropriate cases, test the products for environmental risks which, if unreasonable, can prevent the manufacture and sale of the product. Unlike in the case of the Federal Insecticide, Fungicide, and Rodenticide Act, EPA has the burden of establishing the need for, and the types of, tests required to identify the potential for adverse health and environmental effects of existing and new chemical substances and mixtures.

E. THE CLEAN AIR ACT[62]

The Clean Air Act (CAA) approaches the problem of air pollution through the combination of environmental goals and technology standards. Generally speaking, where the sources for air pollution endangering public health and welfare are numerous and widely distributed and therefore contribute to area-wide air contamination,

EPA must establish nationally uniform ambient air quality standards (a/a/q/s). At the present time, six pollutants are so regulated: particulate matter, sulfur dioxide, carbon monoxide, oxone, nitrogen dioxide, and lead. Following establishment of these standards, the states are required to design the point and area source emission limitations to attain and maintain the standards by dates specified in the Act.

Failure of a state to design an adequate control strategy (as determined by U.S. EPA) results in cut-off of certain highway funds, restrictions on new source growth, and EPA promulgation of a control strategy for the state. New sources cannot be constructed in so-called "non-attainment" areas (i.e., where the a/a/q/s have not been attained) unless they meet especially stringent technology standards (LAER) and their emissions are at least off-set by the reduction of emissions from existing sources. When ambient air quality is better than the national standards (so-called PSD areas), states are required to limit new source growth so as to prevent deterioration of existing ambient air quality beyond the amounts of deterioration permitted by the CAA and EPA regulations. New sources in such areas must install BACT, that is, best available control technology.

With regard to hazardous air emissions, Congress has identified those materials subject to regulation, with the possibility of the addition or deletion of a substance from the statutory list. By dates specified in the legislation, EPA must promulgate emission limitations based on the "maximum degree of emission reduction achievable" for both new and existing sources of these substances. If these technological standards are not adequate to avoid adverse health effects, EPA must later adopt even more stringent standards.

EPA is also responsible for adopting new source performance standards based on best available technology and applicable to all new sources of certain kinds that are considered significant sources of air pollution.

Within the next few years, all significant new and existing sources of air pollution will have to obtain operating permits from the states or EPA, which permits will contain all applicable emission limitations, monitoring, reporting, and other requirements.

The Act also contains more specialized programs to reduce automobile air pollution through the establishment and enforcement of tailpipe emission and fuel formulation standards; to eliminate acid rain through the creation of a marketable allowance system (described above); and to phase out ozone-depleting materials.

F. THE CLEAN WATER ACT[63]

While historically administration of the Clean Air Act has been directed primarily to the achievement of environmental quality standards (i.e., the national ambient air quality standards), the thrust of the effort under the Clean Water Act to date has been largely the imposition of technology standards on new and existing point sources of water pollution though water quality standards are part of the regulatory mechanism. Non-point pollution (e.g., agricultural runoff) has been only marginally impacted, though it remains the most significant source of existing contamination of surface waters. Pollutants are characterized as conventional, nonconventional, and toxic and the degree of pollution reduction required varies from less to more stringent on this basis. New source discharges are more stringently limited than existing ones.

The keystone in the administration of the point source program is the National Pollutant Discharge Elimination System, a permit program administered by the states (or EPA where states have not established federally approvable frameworks), whereby federally established technology standards (expressed as discharge limitations) are written into individual existing and new source permits (with some provisions for waiver or modification in the case of special circumstances). Municipal discharges are subject to special limitations, as are industries that discharge, not directly into the "waters of the United States", but to a municipal system that ultimately so discharges its effluent. Permits to place dredged or fill material into the navigable waters (including adjacent wetlands) are issued by the Corps of Engineers with EPA guidance and veto power.

In addition, the Clean Water Act seeks to attain water quality criteria, particularly for toxics, which are established by the states under EPA guidance. Attainment of these criteria requires, in turn, the allocation among discharges in a watershed of permissible loadings. From this allocation, applicable discharge limits can be written into the permits of existing and new point sources. This approach is premised on the assumption (which has proven correct) that compliance with technology standards by point sources may not by itself attain the clean water goals established by Congress in 1972.

With regard to the particular problem of oil pollution, Congress substantially amended various statutes (including the Clean Water Act) in 1990 to make more effective a system for the prevention and clean-up of oil spills. Oil spill response planning and coordination are emphasized, as are design standards for oil tankers. With regard to the clean-up obligations and cost allocation, Congress modeled the new Oil Pollution Act[64] on various provisions contained in the Comprehensive Environmental Response, Compensation, and Liability Act, CERCLA (described below).

G. THE SAFE DRINKING WATER ACT[65]

First enacted in 1974, the SDWA has been amended no less than five times since then, including important revisions in 1986, which changed the process for establishing health-protective drinking water standards. EPA is required to propose a maximum contaminant level goal (MCLG), which is nonenforceable, for each contaminant that, in the Administrator's judgment, may have adverse health effects and which is known or anticipated to occur in public water systems. Each MCLG must be set at a level at which no known or anticipated adverse health effects occur and which allows for an adequate margin of safety. At the same time as EPA proposes an MCLG, it must also propose a maximum contaminant level (MCL) which is to be as close to the MCLG as is "feasible" (using the best available technology, taking costs into consideration). Alternatively, EPA may require use of treatment techniques in lieu of establishing an MCL if it is not economically or technologically feasible to determine such a level.

The EPA is also required to issue secondary drinking water regulations that specify the maximum contaminant levels necessary to protect public welfare and which deal largely with contaminants affecting drinking water odor and appearance. These standards are not federally enforceable but are issued only as guidelines for the states.

The primary enforcement authority for public water regulation lies with the states, provided they adopt regulations as stringent as the federal ones, develop adequate procedures for enforcement, maintain records, and create plans for providing emergency water supplies. Where EPA finds that a public water system in a state with primary enforcement authority does not comply with applicable regulations, the state must report the steps taken to ensure compliance. Continued failure to comply authorizes the EPA to commence administrative and judicial enforcement actions. Owners and operators of public water systems must also give notice to the persons served by them of failures to comply with MCLs and treatment techniques.

The SDWA grants the power to the states and EPA to issue variances and exceptions in some cases, though their issuance cannot result in an unreasonable health risk. EPA also has emergency powers to issue orders and to sue if a contaminant likely to enter a public drinking water supply system poses a substantial threat to public health. As in so many other environmental statutes, citizens are empowered to sue for violations of the Act.

The SDWA also requires EPA to promulgate regulations for state underground injection control programs to protect underground sources of drinking water. States were to prohibit by December 1977 any underground injection that was not authorized by a state permit. In those cases where a state does not adopt an approvable program, EPA must implement the regulatory regime. For oil and gas injection operations only, states are delegated primary enforcement authority (when they have existing programs for control) without having to meet EPA regulations.

The Act also provides for federal grants for a variety of purposes, including public water system supervision; demonstration programs to develop, implement, and assess critical aquifer protection areas; and state programs for development and implementation of wellhead protection.

H. THE COMPREHENSIVE ENVIRONMENTAL RESPONSE, COMPENSATION, AND LIABILITY ACT[66]

This focuses on the remediation of sites that have been used in the past for the disposal of hazardous substances. Congress has provided money to finance this effort (the "Superfund"). More importantly, present and former owners of the disposal sites, generators of the substances, and transporters (all of which categories are broadly construed by the courts) are liable for the full costs to the government for its clean-up efforts, can be forced by administrative and judicial order to directly incur remedial expenses in cleaning up sites, and may be sued by other persons for their costs of clean-up. Owners, generators, and transporters are also liable to the government as trustee for natural resource damages occasioned by the release of hazardous substances into the environment. There are few defenses to liability. Moreover, where many parties have deposited hazardous substances at a site, each is likely to be found liable for the full amount of the site remediation costs, even if, as is often the case, it contributed only a small part of the total volume of the substances disposed of. Disputes arising under CERCLA generally focus on whether the types of response and remedial actions required in particular cases (that is, the costs of the clean-up) are appropriate and how liability for these costs will be distributed among potentially responsible parties.

I. THE RESOURCE CONSERVATION AND RECOVERY ACT[67]

Establishing a "cradle to the grave" regulatory system, RCRA regulates hazardous waste generators, transporters, and disposal/treatment/storage facilities. Generators are required to use certain types of containers, to label them in a particular way, to maintain certain records and provide information about wastes, and to employ a manifest system that tracks the wastes to their disposal. Transporters must ensure that the manifests are accurate and deliver material in accordance with manifests. Owners and operators of disposal sites must treat, store, and dispose of wastes in a manner consistent with EPA regulations; maintain records; build facilities to meet EPA-specified design standards; monitor for releases of hazardous substances; and take corrective action in the case of a release. A permit system applicable to disposal sites is established by RCRA, as in so many other environmental statutes. Disposal site operators are also subject to various qualification and financial responsibility standards. Land-based disposal techniques are disfavored. EPA's treatment regulations for land disposal are based, in part, on technology standards. One of the more complex and controversial issues in RCRA's administration has, not surprisingly, related to EPA's definition of covered hazardous waste.

J. THE EMERGENCY PLANNING AND COMMUNITY RIGHT-TO-KNOW ACT[68]

States are required to establish committees to develop plans identifying facilities with substantial amounts of hazardous materials and providing for emergency response actions to deal with releases of those substances. The Act also requires public reporting of the nature and characteristics of certain hazardous materials held or used at local facilities.

VII. CONCLUSION

Over the last 22 years, environmental law in the U.S. has evolved from a largely common law base with a few scattered, generally vague, statutes to the point that the volumes containing federal and state environmental statutes, regulations, and agency guidelines of mind-numbing detail and complexity could easily consume the space of a fair-sized public law library. For the practitioner — whether he or she be lawyer, engineer, or scientist — the temptations to specialization with respect to one or a few types of pollution problem are almost overwhelming. For better or ill, as these professionals face the future, they can anticipate that the outpouring of statutes and regulations will continue unabated.

REFERENCES

1. Grad, F., *Treatise on Environmental Law,* 101, 1-4, 1-5, Matthew Bender & Co., New York, 1992.
2. Id.

2a. Id., 1.01, 1-5.

3. Frank, N., *From Criminal Law to Regulation: A Historical Analysis of Health and Safety Law*, Matthew Bender & Co., New York, 1986.
4. Id. at 132.
5. Id.
6. Id. at 139.
7. Craun, G., Ed., *Waterborne Diseases in the United States*, Matthew Bender & Co., New York, 1986, 234.
8. Frank, N., *From Criminal Law to Regulation: A Historical Analysis of Health and Safety Law*, 1986.
9. Id. at 146.
10. Id. at 150-151.
11. Id. at 151.
12. Laitos, Legal institutions and pollution: some intersections between law and history, 15, *Nat. Res. J.*, 423, 433, 1975.
13. Id. at 434.
14. Id. at 444.
15. Id. at 431-433, 448-449.
16. Id. at 437.
17. Reitze, A., A century of air pollution control law: what's worked; what's failed; what might work, 21, *Environ. Law*, 1550, 1991.
18. Plater, Z., Abrams, R., and Goldfarb, W., *Environmental Law and Policy: Nature, Law, and Society*, West Publishing Co., 1992.
19. Thomas, L., Environmental Progress and Challenges: EPA Update, U.S. Environmental Protection Agency, Washington, D.C., 1988.
20. The Pollution Prevention Act of 1990, 42 U.S.C. §§13101-13119.
21. U.S. Constitution, Article I, Section 8, cl. 1 and 3.
22. New York v. United States, 112 S. Ct. 2408. 2419. 2423. 1992.
23. Heart of Atlanta Motel, Inc. v. United States, 379 U.S. 241, 1964.
24. U.S. Constitution, Article VI, Section 8, cl. 2.
25. Pacific Gas and Electric v. California Energy Resources Conservation and Development Commission, 461 U.S. 190, 1983.
26. Ray v. Atlantic Richfield Co., 435 U.S. 151, 1978.
27. City of Philadelphia v. New Jersey, 437 U.S. 617, 1978.
28. Chemical Waste Management, Inc. v. Hunt, 112 S. Ct. 2009, 1992.
29. Maine v. Taylor, 477 U.S. 131, 1986.
30. Pike v. Bruce Church, Inc., 397 U.S. 137, 1970.
31. U.S.C. §§2021b et seq.
32. New York v. United States, 112 U.S. 2408, 1992.

33. U.S. Constitution, Amendment V.
34. Lucas v. South Carolina Coastal Council, 112 S. Ct. 2886, 2893, 1992.
35. U.S.C. §§551 et seq.
36. Sunstein, Constitutionalism after the New Deal, 101, Harv. Law Red., 421, 1987.
37. Oren, Detail and delegation: a study in statutory specificity, 15, *Columbia J. Environ. Law*, 143, 1990.
38. 42 U.S.C. §7604 (Clean Air Act).
39. Stewart, R., The reformation of American administrative law, 80, *Harv. Law Rev.*, 1667, 1975.
40. Regulatory Program of the United States Government, April 1, 1990 — March 31, 1991, U.S. Government Printing Office, Washington, D.C., 1991, p 13-26.
41. 5 U.S.C. §553.
42. Id. 5 §554, 556, and 557.
43. NLRB v. Bell Aerospace Co., 416 U.S. 267. 1974.
44. Mann, O., Polluter-financed environmentally beneficial expenditures: effective use or improper abuse of citizen suits under the Clean Water Act?, 21, *Environ. Law*, 175, 182, 1990.
45. Strauss, P., *An Introduction to Administrative Justice in the United States*, Carolina Academic Press, Durham, North Carolina, 1989.
46. Allen v. United States, 588 F. Supp. 247. 1974, D. Utah.
47. Pierce, R., The role of constitutional and political theory in administrative law, 64, *Tex. Law Rev.*, 469, 1985.
48. 7 U.S.C. §§136-136y.
49. 42 U.S.C. §§110001-11050.
50. 42 U.S.C. §§7661a(b) (3).
51. 7 U.S.C. §§136-136y.
52. 15 U.S.C. §§2601-2671.
53. 21 U.S.C. §348(c) (3) (A).
54. Les v. Reilly, __F.2d__(9th Cir. 1992, 5 Ad. L. 3d 333 (pesticide residues).
55. 42 U.S.C. §§9601-9675.
56. Ackerman and Stewart, Reforming environmental law, 37, *Stanford Law Rev.*, 1333, 1985.
57. Hahn and Hester, Marketable permits: lessons for theory and practice, 16, *Ecol. Law Q.*, 361, 1989.
58. 42 U.S.C. §§4331-4335, 4341-4347.
59. 16 U.S.C. §§1531 et seq.
60. 7 U.S.C. §§136-136y.
61. 15 U.S.C. §§2601-2671.
62. 42 U.S.C. §§7401 et seq.
63. 33 U.S.C. §§1251 et seqq.
64. Pub. L. No. 101-380, 1990.
65. 42 U.S.C. §§300f et seq.
66. 42 U.S.C. §§300f et seq.
67. 42 U.S.C. §§6901-6991K.
68. 42 U.S.C. §§11001-11050.

CHAPTER 2

Environmental Risk Assessment

Irva Hertz-Picciotto

I. INTRODUCTION

Quantitative risk assessment plays a major role in the setting of occupational and environmental standards or exposure limits. It draws on a broad range of disciplines, including toxicology, biostatistics, industrial hygiene, environmental engineering, pathology, genetics, medicine, and epidemiology, to name a few. This chapter introduces the field of risk assessment, its motivation, the approaches typically used, controversies about the accepted methods, and the role for epidemiology.

II. WHAT IS RISK ASSESSMENT?

Risk assessment is the means through which scientific input is provided for guiding regulation of environmental and occupational exposures with potential adverse health effects. In 1983, the National Academy of Sciences summarized risk assessment as "the use of the factual base to define the health effects of exposure of individuals or populations to hazardous materials and situations."[1] This definition reveals the fundamental problem: the "factual base" is generally deficient in that it usually pertains to the wrong species, the wrong route, the wrong dose range, or to an experimental rather than real-world environment. Indeed, estimation of risks for human populations exposed to low doses of environmental chemicals entails seemingly insurmountable problems. For this reason, risk assessment has been viewed with considerable skepticism by many in the scientific community. We return to some of the issues raised by skeptics later in this chapter.

Risk assessment has been distinguished from risk management, described as a separate stage that involves weighing alternative policies and selecting a regulatory action based on both the health risk assessment and other considerations, including technical feasibility, cost, and social and political concerns. The two phases, risk

0-87371-573-X/95/$0.00+$.50

assessment and risk management, are therefore distinct yet linked. They are usually carried out not only by different individuals (risk assessment by scientists, risk management decisions by policy administrators) but often by different institutions, e.g., health departments or hired consultants vs. administrative boards, whose members may include political appointees.

Epidemiologists sometimes serve as "expert witnesses" in legal disputes involving individuals exposed to hazardous agents. In the courtroom, the requirements for legal proof of causality in individuals contrast with scientific criteria for causality in a population.[2] Besides litigation, another direct "real-world" impact of our work is in the regulatory arena, where risk assessment lies at the interface between science and policy. Here, the appropriate criteria for regulating do not necessarily match the criteria for scientific consensus on causality. For example, if the exposure is sufficiently widespread and the consequences serious, then in the spirit of primary prevention, even a moderate amount of evidence may justify regulatory action.

Whereas epidemiologists are always estimating risks, the question asked in risk assessment is more specific. Epidemiologists often address "What is the risk of disease Y in the presence of agent X relative to the risk of disease Y in the absence of X?" The risk assessor asks: "How many excess cases of disease Y will occur in a population of size Z due to exposure to agent X at a dose level D?" Consider an example where outpatient visits for asthma is the outcome and atmospheric SO_2 is the exposure. The epidemiologist might compare the rate of asthma cases on days when SO_2 levels are high to the rate on days when SO_2 levels are low, and compute the ratio of the former to the latter. In contrast, the risk assessor will compute the number of excess cases in a community with a population of a given size, over a period of time (perhaps a year) for all days when SO_2 levels reach certain concentrations, for example, 50 $\mu g/m^3$ or more in ambient air. Either a hypothetical population of one million persons may be used, or the number of cases in a particular community, e.g., the Los Angeles metropolitan area, may be estimated. These two approaches and the questions they answer are different on two counts at least: first, exposure is defined more quantitatively for risk assessment than for most epidemiologic endeavors — not simply presence of agent X or high vs. low levels, but presence at dose D (50 $\mu g/m^3$ in air); second, the outcome is defined exclusively as added risk rather than relative risk (e.g., 450 excess cases per million persons exposed) and is usually related to a time period (e.g., a year or a lifetime). Added risk is defined as the risk in an exposed group *minus* the risk in a referent group, whereas relative risk is defined as the risk in an exposed group *divided by* the risk in a referent group. Synonyms for added risk are absolute risk, excess risk, additive risk, or risk difference; synonyms for relative risk include multiplicative risk or risk ratio.

III. WHY IS RISK ASSESSMENT NEEDED?

The need to conduct health risk assessments reflects the vast increase in the volume of chemicals produced and their unregulated release into the world ecosystem. Federally mandated Toxic Release Inventory (TRI) reports submitted to the U.S.

Environmental Protection Agency (EPA) for 1988 revealed that U.S. industries emitted an estimated 2.4 billion lb. of toxic pollutants into the atmosphere.[3] Of 143 air pollutants measured at sites throughout the U.S., 38% exceeded the health reference level (a level considered "safe" by the U.S. EPA) at one or more sites and 22 exceeded the reference level at more than 25% of the sites studied.[4] These 22 include known carcinogens, reproductive toxins, and neurotoxins. Another 2.1 billion lb. of toxic chemicals were released directly into surface water, land, or underground.[5] The 123 carcinogens on the TRI list accounted for 7% of total releases and transfers. For the majority of chemicals reported, little data are currently available regarding chronic adverse health outcomes, such as reproductive toxicity, mutagenicity, effects on the immune system, neurological impairment, etc. The potential for substantial compromise of health is undeniable.

Situations in which risk assessment is needed for policy decisions are everywhere, including evaluation of risks to employees in a public building with elevated radon levels, residents living near a hazardous waste facility, flight attendants with passive exposure to cigarette smoke, workers engaged in demolition of buildings with asbestos insulation, a community near a proposed municipal incinerator, children attending a school in which the water supply is contaminated, consumers ingesting pesticide residues in food, etc. Risk assessment may also be used for setting priorities. By ranking chemicals according to their potency and the extent of human exposure, agencies may sort through a large number of potential hazards and determine where to focus their regulatory efforts.

IV. THE ROLE OF HEALTH SCIENTISTS

Without risk assessment, the default assumption is frequently that of zero risk. "No risk has been shown" is easily interpreted as "there is no risk", and risk assessment as a way of thinking can guard against this pitfall. As an illustration, consider a pregnant woman in the early 1960s asking advice from her physician regarding the question of alcohol consumption in moderation (e.g., a glass or two of wine per day): the most cautious advice would likely have been "no one has shown any risk". In fact, women were advised that alcohol was harmless and ethanol was even used in a clinical trial to prevent premature labor.[6] Today, of course, the risks associated with alcohol consumption during pregnancy are widely accepted. Clearly, when the database is inadequate or nonexistent, the more appropriate characterization would be "our knowledge is limited", though experts are not always this accurate. It should also be recognized that not all chemicals are harmful.

In response to widespread pollution, regulatory decisions will be made (or not made) based on other criteria if human health considerations are not explicitly injected into the process. An evaluation of occupational standards for chemical exposures[7] revealed that there was little correlation between levels at which adverse effects were reported and the level considered safe, defined as the threshold limit value: "to which nearly all workers may be repeatedly exposed day after day without adverse effect".[8] According to Roach and Rappaport,[7] the primary determinant of such standards, while not explicitly known, appears to have been perceived attainability, which, in turn, is likely to have been related to economic considerations. Since those with financial and organizational backing are frequently the primary influences on policy decisions and since public health is often not the highest priority for those interests, health-based concerns may easily be overlooked. In reality, some industries, when required to find safer alternatives, have actually increased their efficiency. In other instances, the price of safety may be high. However, the costs of working and living in an environment polluted by toxic chemicals are also high. Without risk assessment, formal injection of public health-based criteria into the regulatory process may not occur. For instance, the introduction of lead into gasoline was carried out primarily because no low-level data on health risks were available.

Quantitative risk assessment not only fills a need, but also draws on historical precedent. Although in recent years risk assessment has been most prominent in the area of low-level environmental exposures to chemical and physical agents, informal risk assessment is a tradition in public health, particularly in primary prevention. The following example from a letter published in *Lancet*[9] entitled "Immunising Children Infected with HIV" is typical:

> We discussed possible alternative policies for measles immunisation of HIV-infected children in developing countries . . .
> We followed a hypothetical birth cohort of one million children from 9 months to 5 years of age or AIDS death and examined outcomes for the alternative policies. We assumed that all susceptibles would develop measles and that vaccination levels in the absence of a screening programme would be 50%. Because studies of pediatric HIV infection, clinical measles, and measles immunisation in HIV-infected children have not been done, median estimates for variables used in the analysis were obtained from a Delphi survey of our experts. Estimates for uninfected children were obtained from published studies.
> Our model predicts that adoption of an HIV testing and exclusion policy would prevent 3 measles-vaccine-associated deaths in HIV-infected children . . ."

All the elements of formal risk assessment are contained in this example: assumptions are set forth explicitly; estimates for some parameters are taken from the literature; where no data are available, experts were relied on; finally, the risks are presented as the added number of cases per million children. A more formal exercise of this type[10] projected the effect on mortality if Americans consumed only 30% of calories from fat. Their conclusion was that about 42,000 deaths per year, or about 2% of the annual mortality, would be prevented. Clearly risk assessment, in which hypothetical scenarios are constructed using assumptions that are made explicit, has a firm tradition in public health.

The public demands protection and the decision makers need information that will enable them to respond to these demands with the most scientifically defensible regulatory action. The sentiments of citizens in the state that uses more pesticides than any other were expressed in the Safe Drinking Water and Toxic Enforcement Act of 1986 (Proposition 65),[11] an initiative passed by the California electorate, which began:

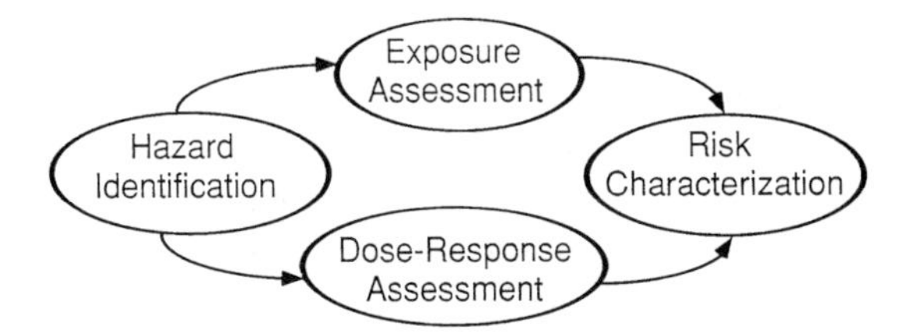

Figure 1 Schematic diagram for the four stages of risk assessment, as discussed in the text.

> The people of California find that hazardous chemicals pose a serious potential threat to their health and well-being, that state government agencies have failed to provide them with adequate protection . . . The people therefore declare their rights: (a) To protect themselves and the water they drink against chemicals that cause cancer, birth defects, or other reproductive harm. (b) To be informed about exposures to chemicals that cause cancer, birth defects or other reproductive harm. (c) To secure strict enforcement of the laws controlling hazardous chemicals and deter actions that threaten public health and safety. (d) To shift the cost of hazardous waste cleanups more onto offenders and less onto law-abiding taxpayers.

While often vocal, the public is also frequently misinformed and wrong in its perceptions. Thus, it is the responsibility of health scientists to expand the state of knowledge and assist in educating the public whenever data are informative. Our challenge is to become clear on the extant data so as to distinguish real dangers from imaginary ones, and to educate the public and policy decision makers on which risks are substantial, which are negligible, and which are, as yet, indeterminate.

In the role of educators, health scientists need to be aware of the ways in which risks are perceived. Research into risk perception reveals that the public perceives those risks that are imposed by others as more objectionable than self-imposed ones (e.g., having contaminated groundwater vs. smoking cigarettes). Citizens are also more disturbed by risks suddenly imposed than by those that have accrued gradually, e.g., an accident or spill vs. increasing levels of air pollutants emitted by automobiles.

V. HOW RISK ASSESSMENTS ARE CONDUCTED

The EPA[12] has outlined four steps in the risk assessment process (Figure 1): hazard identification, exposure assessment, dose-response assessment, and risk characterization. Hazard identification involves determining that a population has been exposed to an agent and that the agent is capable of affecting health adversely. It is largely a qualitative process, relying on existing data as to: (1) the nature of the potential exposure, (2) the likelihood of contact between the human population and the agent, and (3) whether there is a body of evidence from laboratory animal experiments, clinical case studies, and epidemiologic investigations suggesting health risks. The hazard identification phase evaluates two qualitative questions: first, establishing whether exposure has occurred or will occur, and second, determining whether previous research shows reasonable evidence that the exposure may cause adverse effects on human health.

Exposure assessment is the stage where the agent(s) is(are) identified specifically, the route of exposure is determined, and the amount and duration of exposure is quantified. This assessment may involve monitoring air, water, food, or other media, and/or it may involve modeling of exposure over space and time. For instance, if the agents emanate from a point source (e.g., a smokestack, an accident at a nuclear facility) then dispersion models that take account of land topography, meteorologic variables, and the physicochemical properties of the agent or its vectors may be used to obtain population-weighted estimates of exposures. Other considerations include taking account of physiological differences such as breathing rates; using activity profiles that characterize where people spend time (indoors vs. outdoors) to relate either the modeled or measured environmental levels to actual contact between persons and the contaminant; and application of personal, biological, and area monitoring to assess exposures, both current and past.[13]

Complications arise when exposures occur in complex mixtures. In addition, substances may enter the body through multiple routes. A compound released into the air may not only be inspired through the lungs, but may also migrate and enter the body through other media such as food or water, or may deposit on the skin to be absorbed dermally. Further limitations are inherent in exposure assessments based on current environmental measurements or modeling when the disease being evaluated may have been caused by past exposures. Cancer and other conditions that have a long latency are particularly problematic in this regard. Finally, individual behaviors will modify the actual exposure received, and individual physiologic factors will determine actual uptake. For these reasons, the development of biological markers (biomarkers) for exposure and their application in epidemiologic studies has become an active and major area of research.[14,15] Biomarkers are measurements made on individuals that either represent the internal dose, for instance, blood lead, or represent damage that is specific to the agent of interest, e.g., particular DNA adducts measured in proteins, etc. It is hoped that through development of such biomarkers, exposure assessment can be improved, leading to greater reliability of risk assessments.

Dose-response assessment, the third stage of risk assessment, relies on previously published information to describe a relationship between dose and adverse health response, which can then be extrapolated to a (usually) low-level environmental exposure (or to an occupational exposure, which may not be low-level). Different approaches are used for low-dose extrapolation, depending on the nature of the health outcome.

In much of the risk assessment literature, extrapolation for cancer effects is conducted under a nonthreshold assumption, while extrapolation for noncancer effects assumes a threshold. Thus, for endpoints other than cancer, the aim is to determine that dose below which no effect is likely. Animal or human data are used to establish a LOAEL (lowest observed adverse effect level) or NOAEL (no observed adverse effect level). These doses are then multiplied by safety factors to yield a presumed "safe" dose, also referred to as a reference dose or reference level. Commonly used safety factors are: 1/10 for interspecies differences, 1/10 for intraspecies variability, and 1/10 if a LOAEL rather than NOAEL is used.

In contrast to other endpoints, cancer is customarily assumed to follow a nonthreshold mechanism. The rationale is discussed below, but the basic argument is that since the originating events occur at a molecular level in a single cell, extremely low doses are capable of initiating the damage that leads to a full-blown cancer. A variety of mathematical models are available for extrapolating from the observed dose range to the range predicted by the exposure assessment. The choice of models is one area that has generated controversy (see below). Also, when extrapolating between species, a conversion between animal and human doses must be made, either before or after the high- to low-dose extrapolation model is fitted. Dose equivalence can be based on a weight per body weight basis (mg/kg/d), weight per surface area (usually estimated as $mg/kg^{2/3}/d$), a factor that lies between these two ($mg/kg^{3/4}/d$), or cumulative lifetime dose (mg/kg/lifetime). The choice of conversion factor can substantially influence the magnitude of the final risk estimate.[16]

Finally, the risk characterization pulls together the exposure assessment with the dose-response assessment to predict the quantitative risks expected from the given exposure in a defined population. Risk is either presented for a standard exposure level (for instance, the U.S. EPA often uses a "unit dose" of 1 $\mu g/m^3$ in air) or for an estimated or measured exposure. Alternatively, the dose that would produce a one-in-a-million excess risk may be presented. This benchmark is frequently disputed, with some arguing that such risks are trivial and others contending that regulation should take place at even lower levels of risk. In providing final risk estimates, risk characterization involves a multidisciplinary evaluation of all relevant data and an understanding of the limitations in exposure measurement and modeling, in animal toxicity testing, and in clinical and observational epidemiological studies.

Although those producing risk assessments are not always as rigorous as academic scientists may wish, an appropriate risk characterization should not only attempt to communicate limitations, but also specify those points in the risk assessment where uncertainty arises. Optimally, the risk assessor will attempt to assess the magnitude of such uncertainties, either in quantitative terms or in relative terms, so that the risk manager and public can determine what steps introduced the greatest uncertainty. Because risk assessments rely on both data and assumptions, it is important that the risk assessor make explicit what assumptions have been made. Some assumptions relate to internal validity and reliability of data (e.g., whether quality control procedures were followed, how closely independent pathologists' ratings correlated, etc.); others address variability of the parameters estimated from source data; still others originate in decisions for the extrapolation process (e.g., what standard was used for dose equivalence between species, or what mathematical model was selected) and may be based on scientific consensus or considerations specific to the particular agent, route of exposure, or population. Different types of assumptions should be clearly distinguished.

Compared with carcinogenicity assessments, evaluating reproductive toxicity involves several additional levels of complexity. A toxin may be effective in diminishing male reproductive capacity or female reproductive capacity, or in adversely affecting the anatomic or functional development of the embryo, fetus, infant, or child. Unlike cancer, the types of physiologic disruption that can affect these processes are extremely diverse. Reproductive impairment can result from insults to the gonads, the pituitary, the uterus, the seminiferous tubules, etc. The timing for male vs. female reproductive insult can be very different, since sperm development takes place over only a few months, while a female at birth already has all her oocytes. Additionally, extrapolation from animal species raises more complex problems: reproduction in animal species may differ in fundamental ways from human reproduction. For instance: some aspects of development that occur prenatally in humans occur postnatally in some species; the human menstrual cycle bears little resemblance to the rodent estrus cycle; the presence of multiple offspring in each pregnancy leads to the issue of litter effects that do not directly translate to the human situation; resorption sites in rodents, an indication of offspring whose development was aborted early in gestation, does not have a direct parallel in humans. Thus, the issue of animal to human extrapolation for reproductive hazards is far more complex than it is for cancer, where the established mechanism of multiple sequential alterations in DNA and other factors related to regulation of cell growth and division appears to hold across species.

Much of the work in risk assessment has focused on carcinogenicity. Though some attention has been paid to adverse reproductive effects, methodologies for assessing risks from other types of damage, e.g., cardiovascular, immunologic, or neurological impairment, are less developed. Regulatory agencies have developed no formal guidelines or policies, aside from the referense dose approach, for assessing, e.g., long-term chronic neurotoxicity. Yet cancer often will not be the endpoint most amenable to detection through epidemiologic, or even animal experimental, studies, particularly since there is usually a long latency between exposure to an agent and the full-blown clinical manifestation of disease. Note that chemicals causing acute toxicity may also have long-term chronic sequelae as demonstrated by a 2-year follow-up study of agricultural workers with an acute episode of organophosphate poisoning.[17] At this point, methodologies and databases for assessing risks of chronic noncancer adverse health outcomes are needed.

VI. CONTROVERSIES, QUESTIONS, AND UNCERTAINTY

As noted above, in the absence of adequate data, assumptions must substitute for facts, introducing uncertainty. The choice of assumptions may be viewed as a matter of judgment; different scientists may choose different assumptions when faced with the same situation. Also, although risk management is generally a separate process, some aspects of the risk assessment stage may also reflect policy considerations, e.g., the choice of health protective extrapolation models over other models. In this section we discuss controversial issues in the field of risk assessment, some of which have received a great deal of attention and others which are just beginning to emerge.

A. STANDARD ANIMAL TESTING

Over the last few decades, there has been a national program for laboratory testing of chemicals for carcinogenicity, administered first by the National Cancer Institute and then by the National Institute of Environmental Health Sciences. Hundreds of chemicals have been tested in several species, creating an impressive and useful database for risk assessors.[18] However, the applicability of these bioassay results to human conditions has been questioned.[19-21] Critics have argued that the testing of animals at "maximum tolerated doses" (MTDs) and at half the MTD causes inflammation and cell proliferation which does not occur at low exposures and that elevated rates of mitogenesis increase the opportunities for spontaneous mutations, hence contributing to carcinogenic development. Advocates of the testing program point out that: (1) 90% of the carcinogens identified by the National Toxicology Program induced tumors in organs that showed no evidence of cellular toxicity; (2) toxicity, while frequently observed at the MTD, is usually not observed at half MTD, even though increased tumor incidence usually is; and (3) toxicity or cell proliferation is often produced without accompanying carcinogenicity, evidence that mitogenesis is not sufficient by itself to produce carcinogenicity.[22-26] Notably, risk predictions from rodent studies are often reasonably close to observed cancer mortality in epidemiologic studies of populations exposed to the same agent.[27-30] For instance, risks observed among workers exposed to ethylene oxide during its manufacture were remarkably close to the risks extrapolated from rats exposed via inhalation.[29,31] Additionally, the list of carcinogens identified first in animals and subsequently shown to cause cancer in humans is growing. On the other hand, some chemicals known to cause cancer in humans have never been confirmed as animal carcinogens (e.g., arsenic), and there have been quantitative discrepancies, for instance, between the animal-based risk estimates and those based on an occupational study of cadmium-exposed workers.[29, 32]

Interestingly, the critics of animal-based risk assessment rarely point out another problem with the standard cancer bioassay: testing chemicals one by one presents unrealistic scenarios in which multiple interactions cannot be observed. A creative set of experiments involving complex mixtures was conducted by a group at the U.S. National Institute of Environmental Health Sciences.[33] In one of these, mice were given a complex mixture of 25 common groundwater contaminants reported by EPA to be frequently found near toxic waste dumps.[34] The exposure levels were kept at concentrations found in environmental samples and results showed suppression of immune function, including granulocyte-macrophage colony formation, hematopatetic stem cells, and antibody-forming cells. Altered resistance to challenge from an infectious agent was also observed. Subclinical effects may be a more plausible health response than the more typically feared cancers when large populations are exposed to low levels of complex mixtures. Such scenarios deserve greater attention.

While standard animal bioassays are not without limitations, they continue to serve as an important advance warning system, particularly for carcinogenicity. Further work investigating their predictiveness for human exposure situations should be on the agenda.

B. THRESHOLDS FOR NONCANCER EFFECTS

For other than acute effects, the assumption and identification of a threshold may be problematic. For instance, 15 years ago, occupational exposures leading to less than 120 μg/dl of lead in the blood were considered to be safe.[35] Today, effects on neurodevelopment in children have been observed at blood levels one tenth of that,[36] and at 2 to 15 μg/dl, a measurable effect on blood pressure has been observed.[37] Associations with low birthweight and preterm delivery have also been observed at less than 40 μg/dl.[38,39] A recent paper has shown decrements in renal function at low exposures, mainly under 20 μg/dl.[40] Adverse health effects have clearly been observed in populations exposed only to "low" environmental levels of lead; a clear threshold remains to be identified. It seems unlikely that lead would be unique in its capacity to cause damage at exposures once considered safe.

C. THRESHOLDS FOR CANCER

For cancer effects, the usual assumption is that there is no threshold. The basis for this assumption is the single cell origin of cancer. Other types of damage only begin to constitute a health problem when entire tissues are incapacitated, beyond the point where either regeneration of functional units or development of compensatory mechanisms are possible. In the last decade, the revolution in molecular biology has elucidated the role of specific genes in the induction of cancer. Researchers have identified oncogenes that stimulate cell growth, suppressor genes that normally inhibit proliferation of cells, and genes involved in DNA repair, DNA synthesis, and integrity of the mitotic spindle.[41] Unlike most other types of toxic damage, the pathway for carcinogenicity involves multiple sequential alterations in DNA of a single cell and its progeny. Thus, it is argued that no matter how small the dose, there is a finite probability of activating an oncogene, or, in an initiated cell, of inactivating a tumor suppressor gene responsible for controlling mitosis, thereby resulting in uncontrolled proliferation.[42,43] Objections to the nonthreshold assumption relate both to the nonlinear kinetics by which some procarcinogens are metabolized and to the action of nongenotoxic chemicals that test as carcinogens in animal bioassays. Objections to the use of standard bioassays have also been presented as arguments against the nonthreshold assumption. The critics contend that increased mitogenesis is unlikely to occur at low doses and is only a function of the near-toxic levels of exposure.[19,44] At this point, direct evidence regarding rates of cell proliferation at high vs. low doses is lacking. Nevertheless, the difficulty in establishing a threshold for any particular compound has presented a barrier to implementation of alternative approaches.

D. MODEL CHOICES

Under the nonthreshold assumption, a variety of mathematical models have been developed for extrapolating from an observed dose-response curve at high exposures to a predicted dose-response relationship at low exposures. For extrapolating from

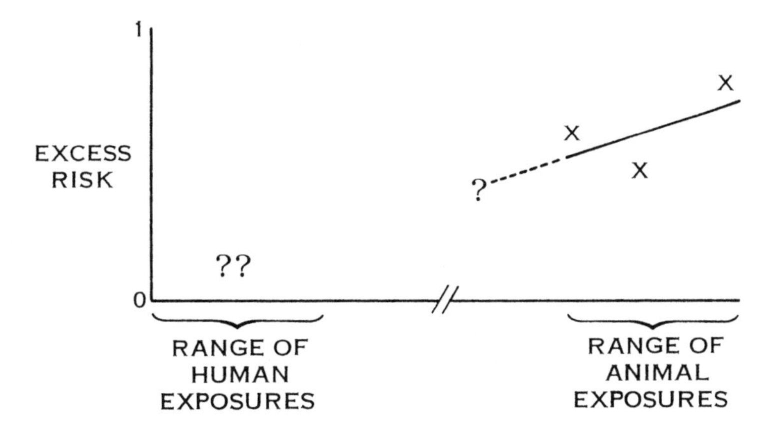

Figure 2 Schematic representation of the dose-response extrapolation to low-level exposures from studies conducted at higher exposure levels.

animal bioassays, these include, among others, the multistage, probit, logit, multi-hit, and Weibull models.[45,46] The choice of model can have a large impact on the final risk estimate, particularly if the human exposures of interest are as much as 100 or 1000 times lower than the doses used in the animal experiments (eg., see Figure 2). Some authors have reported that discrepancies produced from the use of different models are 1000-fold or greater. Such differences are not surprising, since one can postulate almost any shape to a dose-response curve that is fitted to only a few points and extrapolated far outside the range of observed data. Nevertheless, if some reasonable *a priori* assumption is made concerning the shape of the curve in the region of actual or projected exposure, then the range of predicted risks can be narrowed considerably. For instance, a common assumption is that at low doses, the response increases linearly with additional exposures. The linearity assumption has a number of advantages, including: (1) linear models are robust to errors in measurement (if exposures have been under- or over-estimated, the degree of distortion is lower using a linear model than a curved one); (2) in many cases, the assumption of linearity provides a health "conservative" (protective) estimate of risk, that is, it may serve as an upper-bound; (3) the proposition that nonreversible DNA damage occurs in direct proportion to the amount of carcinogen contacting the tissue is biologically plausible. By making an assumption of low-dose linearity, most of the widely divergent risk extrapolation models will be excluded. In practice, most risk assessments implement some version of the multistage model, which provides a dose-response curve that is linear at low doses. The observed data simply cannot, in fact, distinguish among models, and the choice must be based on considerations other than "statistical goodness of fit".

Although most of these models have been developed to extrapolate from experimental animal data, either animal or human data can be used for both noncancer and cancer risk assessments. As the International Agency for Research on Cancer succinctly stated, ". . . in the absence of adequate data in humans it is reasonable, for practical purposes, to regard chemicals for which there is sufficient evidence of carcinogenicity (i.e., a causal association) in animals as if they presented a carcinogenic risk for humans".[47] If human data are not available, then there is no choice but to use the animal data. When both are available, and the human data are of sufficient quality, these are generally preferable over animal data as a basis for extrapolation.[48] For many agents, it is not an "either/or" situation; data from several species can be used. At times, however, the data are discordant, either qualitatively or quantitatively. Standard practice is to choose the most sensitive species as the basis for a risk assessment.

E. ADVANTAGES OF EPIDEMIOLOGIC DATA

Quantitative risk assessment has put great emphasis on animal experimental data. This practice has been defended on the grounds that epidemiologic data involve uncertain measures of exposures as compared with high-quality, well-controlled dosing of animals, and that the experimental model is free from biases such as confounding. In reality the net advantage of using human data is far greater than the disadvantages.[48]

First, the magnitude of error is likely to be greater using animal data because interspecies differences incur greater uncertainty than the major sources of uncertainty in epidemiologic studies.[48,49] The interspecies differences include absorption rates, metabolic pathways, and rates for activation or detoxification and elimination, as well as the possibility that the target site does not even exist in humans. These differences can entail differences of several orders of magnitude.

Another advantage of human data is that the range of extrapolation is usually smaller. Available data frequently come from occupational studies. Though exposures in industrial or agricultural settings can be relatively high, they are often still lower than the doses used in animal experiments (see Figure 2). Thus, in most cases, using human data from an occupational setting reduces the range of extrapolation to environmental levels as compared with the range required by extrapolation from animal data. In rare cases, environmental epidemiologic studies may be large enough, or the adverse effects strong enough, to be detected by observational studies. Lead has been shown to cause neurodevelopmental deficits in children at exposure levels that were commonplace in the U.S. in the 1970s. In these instances, the range of extrapolation may be minimal, depending on the population of interest.

A third reason to favor human data over animal data is that the genetic diversity and the variability in other endogenous or host factors in the human population will be better represented in another human study than in an animal study.[48] Because susceptibility to disease will differ according to these factors, the controlled experimental situation in which a single strain in each of one or two species is tested has less generalizability than any human study.

F. INTERSPECIES EXTRAPOLATION

When human data are scant or lacking entirely, extrapolation from animal species is unavoidable. As noted above, species differences can be large and difficult to assess. How valid are these extrapolations?

The problem of assessing validity is in some ways insurmountable. There may never be a way to know whether the risks estimated at low doses are "correct". Some researchers have used studies of humans who experienced exposures intermediate between the experimental doses used on animals, and the low levels to which risks were being extrapolated. Risks predicted by applying the low-dose extrapolation model to an intermediate level of exposure were compared with actual cancer mortality. Comparisons between observed and expected cancer risk often showed remarkable consistency.[28-31]

Other researchers have compared potencies calculated from animal-based risk assessments with potency calculations based on human data.[27] Both consistent and discordant results have been observed. However, it is important to keep in mind the scale on which consistency ought to be evaluated. Risk assessment does not aim to estimate exact numbers of deaths, but rather ball-park figures. If estimates are within three- to fivefold of the true risks, then the process may be viewed as working reasonably well. If estimates are usually tenfold or more too high or too low, then the process needs revising. A fairly thorough comparison of animal- and human-based risk assessments showed a range in which about half the potency values were within fivefold of each other, and the other half were off by a factor of between five- and 100-fold.[27] No clear pattern emerged in terms of what types of chemicals showed more interspecies discordance. In addition, for about half the chemicals evaluated, animal-based risk estimates were higher than human-based estimates, and for about half, the human-based estimates were higher than the animal-based ones. Thus, current methods for extrapolating from animal bioassays do not appear to be systematically over- or under-estimating human risks.

Finally, Tomatis et al.[50] showed that among known human carcinogens that have been adequately tested in animals, the site of human cancer is also a site of animal carcinogenicity in virtually all cases. The most common sites were the lung, the blood-forming organs, and the bladder.

VI. CONCLUSION

Quantitative risk assessment bridges the gap between environmental policy and scientific data regarding health risks. It also serves as an impetus for the development of experimental science in the direction of relevant technology that can help answer the question being posed by concerned citizens: Is it safe? Risk assessment is a growing field that is continuously undergoing re-evaluation of methods and assumptions. In the future, some controversies may be resolved with additional data and with studies of higher quality, while others will likely remain with us for some time. Epidemiologic studies with improved measures of exposure are needed to reduce uncertainty in risk extrapolation. Investigations that combine molecular biology with epidemiology are an active area that will certainly play a role in bringing about better exposure assessments, including measures of internal dose. Other areas needing attention are methods to integrate human and animal data, standardized criteria for evaluating appropriateness of data to be used in risk assessment,[48] greater attention to areas other than carcinogenicity, including neurotoxicity and immunotoxicity, and further investigations into the impact of multiple exposures and ways to study them.

REFERENCES

1. National Research Council, National Academy of Sciences, Risk Assessment in the Federal Government: Managing the Process, National Academy Press, Washington, D.C., 1983.
2. Christoffel, T. and Teret, S. P., Epidemiology and the law: courts and confidence intervals, *Am. J. Public Health*, 81, 1661, 1991.
3. U.S. Environmental Protection Agency, Toxics in the Community, National and Local Perspectives, The 1988 Toxics Release Inventory National Report, (TS-779), Publication No. EPA-560/4-90-017, U.S. Environmental Protection Agency, Washington, D.C., 1990.
4. Hassett-Sipple, B., Cote, I., and Vandenberg, J., Toxic air pollutants and noncancer health risks — United States, *MMWR*, 40, 278, 1991.
5. U.S. Environmental Protection Agency, Toxics in the Community, National and Local Perspectives, The 1989 Toxics Release Inventory National Report, (TS-779), EPA 560/4-91-014, U.S. Environmental Protection Agency, Washington, D.C., 1991.
6. Zlatnick, F. J. and Fuchs, F., A controlled study of ethanol in threatened premature labor, *Am. J. Obstet. Gynecol.*, 112, 610, 1972.
7. Roach, S. A. and Rappaport, S. M., But they are not thresholds: a critical analysis of the documentation of threshold limit values, *Am. J. Ind. Med.*, 17, 727, 1990.
8. American College of Governmental Industrial Hygienists, Threshold Limit Values and Biological Exposure Indices for 1988-1989, Cincinnati, Ohio, 1988.
9. Onorato, I. M., Jones, T. S., and Orenstein, W. A., Immunising children infected with HIV (Editorial), *Lancet*, 1, 354, 1988.
10. Browner, W. S., Westernhouse, J., and Tice, J. A., What if Americans ate less fat?, *JAMA*, 265, 3285, 1991.
11. California Safe Drinking Water and Toxic Enforcement Act of 1986, California Health and Safety Code, Section 25249 *et seq.*; Title 22, California Code of Regulations, Section 12000 *et seq.*, 1990.
12. U.S. Environmental Protection Agency, Guidelines for cancer risk assessment, *Fed. Register*, 51, 33992, 1986.
13. Rappaport, S. M. and Smith, T. J., Eds., *Exposure Assessment for Epidemiology and Hazard Control*, Lewis Publishers, Chelsea, Michigan, 1991.
14. Hulka, B. S., Wilcosky, T. C., and Griffith, J. D., *Biological Markers in Epidemiology*, Oxford University Press, New York, 1990.
15. Vine, M. F. and Hulka, B. S., Biological markers of exposure, chap. 6 in this volume.
16. U.S. Environmental Protection Agency, Draft report: a cross-species scaling factor for carcinogen with assessment based on equivalence of $mg/kg^{3/4}/day$, *Fed. Register*, 57, 24152, 1992.

17. Rosenstock, L., Keifer, M., Daniell, W. E., McConnell, R., Claypoole, K., and the Pesticide Health Effects Study Group, Chronic central nervous system effects of acute organophosphate pesticide intoxication, *Lancet*, 338, 223, 1991.
18. Huff, J. E., McConnell, E. E., Haseman, J. K., Boorman, G. A., Eustis, S. L., Schwetz, B. A., Rao, G. N., Jameson, C. W., Hart, L. G., and Rall, D. P., Carcinogenesis studies: results of 398 experiments on 104 chemicals from the U.S. National Toxicology Program, *Ann. N.Y. Acad. Sci.*, 543, 1, 1988.
19. Ames, B. N. and Gold, L. W., Too many rodent carcinogens: mitogenesis increases mutagenesis, *Science*, 249, 970, 1990.
20. Ames, B. N. and Gold, L. S., Carcinogens and human health: Part 1 (Letter), *Science*, 250, 1645, 1990.
21. Ames, B. N. and Gold, L. S., Carcinogenesis mechanisms: the debate continues (Letter), *Science*, 252, 902, 1991.
22. Infante, P. F., Prevention versus chemophobia: a defence of rodent carcinogenicity tests, *Lancet*, 337, 538, 1991.
23. Perera, F. P., Carcinogens and human health: Part 1 (Letter), *Science*, 250, 1644, 1990.
24. Huff, J. E. and Haseman, J. K., Exposure to certain pesticides may pose real carcinogenic risk, *Chem. Eng. News*, 69, 33, 1991.
25. Rall, D. P., Carcinogens and human health: Part 2 (Letter), *Science*, 251, 10, 1991.
26. Cogliano, V. J., Farland, W. H., Preuss, P. W., Wiltse, J. A., Rhomberg, L. R., Chen, C. W., Mass, M. J., Nosnow, S., White, P. D., Parker, J. C., and Wuerthele, S. M., Carcinogens and human health: Part 3 (Letter), *Science*, 251, 606, 1991.
27. Allen, B. C., Crump, K. S., and Shipp, A. M., Correlation between carcinogenic potency of chemicals in animals and humans, *Risk Anal.*, 8(4), 531, 1988.
28. Hertz-Picciotto, I., Gravitz, N., and Neutra, R., How do cancer risks predicted from animal bioassays compare with the epidemiologic evidence? The case of ethylene dibromide, *Risk Anal.*, 8, 205, 1988.
29. Hertz-Picciotto, I., Neutra, R. R., and Collins, J. F., Evidence for carcinogenicity of ethylene oxide, *JAMA*, 257, 2290, 1987.
30. Alexeeff, G. V. and Hertz-Picciotto, I., Consideration of species concordance and pharmacokinetics in a risk assessment of methylene chloride, *Toxicologist*, 10(1), 351, 1990.
31. Hertz-Picciotto, I. and Neutra, R. R., Resolving discrepancies among studies: the influence of dose on effect size, *Epidemiology*, 5, 156, 1994.
32. Hertz-Picciotto, I. and Hu, Suh-Woan, Contribution of cadmium in cigarettes to lung cancer: an evaluation of risk assessment methodologies, *Arch. Environ. Health*, 49, 297, 1994.
33. Yang, R. S. H., Hong, H. L., and Boorman, G. A., Toxicology of chemical mixtures: experimental approaches, underlying concepts, and some results, *Toxicol. Lett.*, 49, 183, 1989.
34. Germolec, D. R., Yang, R. S. H., Ackermann, M. F., Rosenthal, G. J., Boorman, G. A., Blair, P., and Luster, M. I., Toxicology studies of a chemical mixture of 25 groundwater contaminants. II. Immunosuppression in $B6C3F_1$ mice, *Fundam. Appl. Toxicol.*, 13, 377, 1989.
35. Occupational Safety and Health Administration, U.S. Department of Labor, Occupational exposure to lead, *Fed. Register*, 43, 54353, 1978.
36. Agency for Toxic Substances and Disease Registry, The nature and extent of lead poisoning in children in the United States, A report to Congress, Atlanta, GA, Department of Health and Human Services, Public Health Service, Atlanta, GA, 1988.

37. Sharp, D. S., Osterloh, J., Becker, C. E., Bernard, B., Smith, A. H., Fisher, J. M., Syme, S. L., Holman, B. L., and Johnston, T., Blood pressure and blood lead concentration in bus drivers, *Environ. Health Perspect.*, 78, 131, 1988.
38. McMichael, A. J., Vimpani, G. V., Robertson, E. F., Baghurst, P. A., and Clark, P. D., The Port Pirie cohort study: maternal blood lead and pregnancy outcome, *J. Epidemiol. Community Health*, 40, 18, 1986.
39. Bellinger, D., Leviton, A., Rabinowitz, M., Allred, E., Needleman, H., and Schoenbaum, S., Weight gain and maturity in fetuses exposed to low levels of lead, *Environ. Res.*, 54, 151, 1991.
40. Staessen, J. A., Lauwerys, R. R., Buchet, J. P., Bulpitt, C. J., Rondia, D., Vanrenterghem, Y., Amery, A., and the Cadmibel Study Group, Impairment of renal function with increasing blood lead concentrations in the general population, *N. Engl. J. Med.*, 327, 151, 1992.
41. Nowell, P. C., How many human cancer genes?, *J. Natl. Cancer Inst.*, 83(15), 1061, 1991.
42. Weinberg, R. A., Tumor suppressor genes, *Science*, 254, 1138, 1991.
43. Shields, P. G. and Harris, C. C., Molecular epidemiology and the genetics of environmental cancer, *JAMA*, 266(5), 681, 1991.
44. Cohen, S. M. and Ellwein, L. B., Cell profileration in carcinogenesis, *Science*, 249, 1007, 1990.
45. Van Ryzin, J., The assessment of low-dose carcinogenicity, *Biometrics Suppl. Curr. Topics Biostat. Epidemiol.* 130, 1982.
46. Van Ryzin, J., Quantitative risk assessment, *J. Occup. Med.*, 22, 321, 1980.
47. International Agency for Research on Cancer, *Monographs on the Evaluation of the Carcinogenic Risk of Chemicals to Humans,* Suppl. 1, *Chemicals and Industrial Processes Associated with Cancer in Humans*, International Agency for Research on Cancer, Lyon, 1979.
48. Hertz-Picciotto, I., Epidemiology and quantitative risk assessment: A bridge from science to policy, *Amer. J. Pub. Health*, 85, 484, 1995.
49. Smith, A. H., Epidemiologic input to environmental risk assessment, *Arch. Environ. Health*, 43, 124, 1988.
50. Tomatis, L., Aitio, A., Wilbourn, W., and Shuker, L., Human carcinogens so far identified, *Jpn. J. Cancer Res.*, 80, 795, 1989.

CHAPTER 3

Association and Causation in Environmental Epidemiology

Neal D. Traven, Evelyn O. Talbott, and Erick K. Ishii

How do we decide whether an observed association represents a causal relationship?

I. INTRODUCTION

Environmental regulatory agencies frequently depend on information derived from epidemiologic studies in deciding on proposed occupational and nonoccupational exposure limits. In addition, results of epidemiologic investigations are often used for prediction of potential future disease in exposed populations and attribution of risk in litigations. Careful evaluation of the combined evidence from epidemiologic and other research helps to gauge the likelihood of an association or a causal relationship between a specific risk factor and a specific disease.

Investigating disease causality through natural or designed experiments is reasonably straightforward. However, in the much more common situations where experimental data do not or cannot exist, we rely instead on population-based observational studies conducted among free-living humans. It is far more difficult to ascertain causality directly from retrospective or cross-sectional studies, yet that remains the goal of environmental epidemiology research. To address this challenge, several thinkers have proposed general criteria for drawing causal inferences from such study designs. In this chapter, we review several formulations of the conceptual framework underlying the process of epidemiologic analyses.

It is important to understand that our scientific knowledge is continually evolving. In other words, what we say is sufficient evidence to indicate causality today may be inadequate, or inappropriate, in the future. Support for this assertion may be found by briefly reviewing the history of criteria for causal inference in epidemiological studies.

0-87371-573-X/95/$0.00+$.50

Table 1 Henle-Koch Postulates (1840)

1. The parasite occurs in every case of the disease in question and under circumstances which can account for the pathological changes and clinical course of the disease.
2. It occurs in no other disease as a fortuitous and nonpathogenic parasite.
3. After being fully isolated from the body and repeatedly grown in pure culture, it can induce the disease anew.

Derived from Evans, A. S., *Yale J. Biol. Med.*, 49, 175, 1976.

II. BACKGROUND

Perhaps the most widely known formal conceptualization of epidemiologic analysis, the Bradford Hill postulates, arose from the long-running debate over a question first posed nearly 50 years ago: is there a causal relationship between cigarette smoking and carcinoma of the lung? Early studies by Dorn[1] and later by Doll and Hill[2] stimulated much controversy.

Similar criteria for inferring causality, however, have existed for at least a century and a half. Historically, that early set of causality postulates was developed in 1840 by Swiss anatomist Jakob Henle and elaborated upon by his student Robert Koch, the famed German physician, pathologist, and bacteriologist. Their combined efforts, presented in public lectures by Koch in 1884 and 1890, are known as either the Henle-Koch or Koch postulates (see Table 1).

Even as the Henle-Koch postulates were being promulgated, it was clear that their strict, overly deterministic rules would limit their utility.[3] For example, researchers were unable to transmit some diseases (e.g., typhoid fever, leprosy) to experimental hosts, and unable to isolate and culture any infective agents at all in other diseases. The asymptomatic carrier state violates the second Koch postulate. These limitations resulted, for the most part, from the era's limited scientific technology, incapable of dealing with nonbacterial diseases. The discovery of viruses in 1930 further demonstrated the limitations of the Henle-Koch postulates. American virologist Thomas Rivers showed that the Henle-Koch postulates literally cannot be applied to viruses, which require living tissue for propagation and cannot be grown in pure cultures.

In the 1950s, Robert J. Huebner incorporated Henle-Koch and Rivers in his "Bill of Rights for Prevalent Viruses".[3] His intent was to apply reasoning arising from legal scholarship to scientific research, trying thereby to avoid subjecting microorganisms to guilt by association. Furthermore, he stressed that his suggestions should be regarded not as "postulates" but merely as useful guidelines.

In the last several decades, continued technological innovations, such as electron microscopy, molecular genetics, and immunology, have revealed disease to be a complex interaction among broadly defined agent, host, and environment. As the intricacy of these multifactorial conditions is uncovered, Henle-Koch and its descendants are less and less relevant to the concept of causation.

In 1968, virologist Werner Henle (Jakob's grandson), his wife Gertrude, and German physician Volker Diel claimed that Epstein-Barr virus, or a close relative, caused infectious mononucleosis. They used seroepidemiologic techniques rather than isolating a causative agent or reproducing the disease anew in an animal model.

Table 2 Elements of Immunological Proof of Causation

1. Antibody to the agent is regularly absent prior to the disease and to exposure to the agent (i.e., before the incubation period).
2. Antibody to the agent regularly appears during illness and includes both IgG- and IgM-type antibodies.
3. The presence of antibody to the agent indicates immunity to the clinical disease associated with primary infection of the agent.
4. The absence of the antibody to the agent indicates susceptibility to both infection and the disease produced by the agent.
5. Antibody to no other agent should be similarly associated with the disease unless it is a cofactor in its production.

Derived from Evans, A. S., *Yale J. Biol. Med.*, 49, 175, 1976.

Table 3 Surgeon General's Report (1964)

1. The consistency of the association
2. The strength of association
3. The specificity of the association
4. The temporal relationship of the association
5. The coherence of the association

Derived from Surgeon General, Advisory Committee of the U.S. Public Health Service, PHS Pub. No. 1103, Public Health Service, Washington, D.C., 1964.

Table 4 Bradford Hill Postulates (1965)

1. Strength of association
2. Consistency
3. Specificity
4. Temporality
5. Biological gradient
6. Plausibility
7. Coherence
8. Experiment
9. Analogy

Derived from Hill, A. B., *Proc. R. Soc. Med.*, 58, 295, 1965.

Such studies have resulted in the promulgation of immunological criteria of causation[3] (see Table 2).

While infectious disease research became ever more complex, epidemiology also turned its attention to noninfectious chronic diseases. Clearly, the Henle-Koch postulates are inappropriate to chronic diseases, which have become the primary concern of environmental epidemiology. Recognizing this fact, the 1964 Surgeon General's Report on Smoking included a set of causality criteria[4] (see Table 3).

As noted at the beginning of this section, the Bradford Hill criteria[5] emerged from the extended debate over the causal relationship between cigarette smoking and lung cancer (see Table 4). Supporters of the tobacco industry and others have criticized study design, sampling, research protocols, results, and conclusions of the research; this scrutiny has encouraged epidemiologic researchers to delineate the philosophical and logical underpinnings of causality research.

Table 5 Alfred Evans Criteria for Causation: A Unified Concept

1. *Prevalence* of the disease should be significantly higher in those exposed to the putative cause than in case controls not so exposed[a].
2. *Exposure* to the putative cause should be present more commonly in those with the disease than in controls without the disease when all risk factors are held constant.
3. *Incidence* of the disease should be significantly higher in those exposed to the putative cause than in those not so exposed, as shown in prospective studies.
4. *Temporally*, the disease should follow exposure to the putative agent with a distribution of incubation period on a bell-shaped curve.
5. A *spectrum* of host responses following exposure to the putative agent along a logical biologic gradient from mild to severe.
6. A *measurable host response* following exposure to the putative cause should regularly appear in those lacking this before exposure (i.e., antibody, cancer cells) or should increase in magnitude if present before exposure; this pattern should occur in persons so exposed.
7. *Experimental reproduction* of the disease should occur in higher incidence in animals or humans appropriately exposed to the putative cause than in those not so exposed; this exposure may be deliberate in volunteers, experimentally induced in the laboratory, or demonstrated in a controlled regulation of natural exposure.
8. *Elimination or modification* of the putative cause or vector carrying it should decrease the incidence of disease (control of pollution water or smoke or removal of the specific agent).
9. *Prevention or modification* of the host's response on exposure to the putative cause should decrease or eliminate the disease (immunization, drug to lower cholesterol, specific lymphocyte transfer factor in cancer).
10. The whole should make biologic and epidemiologic sense.

[a] The putative cause may exist in the external environment or in a defect in host response.

In addition to the Bradford Hill criteria, seroepidemiologist Alfred Evans has proposed a set of postulates[3] (see Table 5) aimed at updating the Henle-Koch postulates to account for advancements in technology and scientific information. While the orientations of the Bradford Hill and Evans criteria differ, there are substantial similarities between them. Because they were explicitly developed in response to a kind of environmental epidemiology issue, we will examine the Bradford Hill criteria as they pertain to a specific current problem in environmental epidemiology.

III. DISCUSSION

The Bradford Hill postulates are expressed in broad and nonspecific terms, so as to be generalizable to a wide variety of research questions. Before examining a specific problem, we briefly describe several of the individual postulates as they relate to environmental epidemiology.

Specificity: Except for a few simple infectious diseases, asbestosis, and perhaps mesothelioma, there are no one-to-one relationships between agent and disease. Most diseases have multifactorial etiology, and those conditions associated with environmental factors are particularly complex in their risk factor associations.

Temporality: Though it is intuitively obvious that a causal factor or exposure must precede the disease it causes, this formulation is too simplistic. Interaction between exposure to the agent and the potential host's biological system (DNA repair, enzyme activation/inactivation, and the like) must be taken into consideration. The effects of dose periodicity, cumulative dosage effects, clearance, and metabolism of the putative cause may also influence the temporal relationship between exposure and disease.

Biological gradient (dose-response): If seen, this relationship strengthens the likelihood of causality. But does absence of dose-response weaken the possibility of a causal relationship? Not necessarily — with a threshold effect, one might observe a dichotomous (all or none) outcome. In some situations, the dose-response relationship is nonlinear; specific levels of exposure, but not exposures that are smaller or larger, may be associated with environmental disease.

Plausibility and analogy: As scientific knowledge increases and progresses, novel relationships among biological mechanisms are constantly being demonstrated. That which lacks plausibility, or for which there is no appropriate analogy, today might be quite plausible tomorrow.

Experiment: Ethical considerations preclude human experimentation in all but the most highly controlled settings, with only the least dangerous of materials. While animal models exist for many exposures, interspecies differences (known and unknown) limit their utility and applicability. For instance, exposure to levels of dioxin lethal to large animals results in little more than chloracne in accidental human exposures.

Along with assessing the causality implications of an epidemiologic investigation, it is vital to consider the overall characteristics of the study — data quality, appropriateness of the study design and data collection procedures, statistical power, and sample size. In environmental epidemiologic studies, other factors to take note of include the procedures used to estimate and manage variability in genetic susceptibility, confounding factors, and risk factor interactions. One should also observe how the researchers handle and discuss potential biases, and measure exposures and outcomes.

Furthermore, as previously stated, we must recognize that causality is a continuum and not an all-or-nothing issue. New discoveries may uncover new causal relationships. This constant change does not mean that causal criteria can be ignored, but rather they should be used as a general guide underlying the determination of a causal relationship.

IV. CASE EXAMPLE

Is there a causal association between electromagnetic fields and childhood leukemia?

Strength of an association is the magnitude of the disease rate in exposed persons compared with unexposed individuals. But there is essentially no such thing as a population unexposed to electromagnetic fields (EMF); there are only degrees of exposure. Associations between levels of exposure to EMF and the development of childhood leukemia are weak, and many of the studies demonstrate no association between relatively high exposure and disease.

Studies on EMF do not seem to be *consistent*. In part, this lack of consistency could be due to the lack of standardization in measurement methods (other than Wertheimer and Leeper's wire configuration). In contrast, calibrated and standardized personal dosimeters are readily available for studies of aerosols, noise, etc. Could this inconsistency be due to unmeasured contaminants in the environment causing the leukemias? The studies may support such a conclusion.

Specificity of an association is a measure of its uniqueness. For example, nothing other than exposure to asbestos appears to be causally related to mesothelioma, and asbestos has not been strongly associated with other neoplasms. By contrast, several forms of adult and childhood leukemias appear to be associated with exposure to benzene, ionizing radiation, and certain viruses. These exposures, in turn, have also been shown to be associated with central nervous system (CNS) lesions. This lack of specificity mitigates against a causal explanation.

In assessing the *temporality* of the association, we need to know that exposure to EMF preceded the onset of the leukemia. Since we are always exposed to EMF, it is reasonable to assume that some exposure preceded the disease. But was that exposure sufficient to cause the disease? Also, it is impossible to know the actual time of onset of the disease. Thus, there is little that can be concluded about the temporal association of EMF and childhood leukemia.

In 1979, Wertheimer and Leeper[6] examined electrical wire configurations in relation to childhood cancer. Though they found a *dose-response relationship*, it is unclear what wire configurations really mean; they might be nothing more than a confounding or even extraneous factor. For example, areas with high wire configuration tend to have transient residents, so wire configuration might be little more than a surrogate or confounder for socioeconomic status. This study ignored the important factor of traffic density (automobile exhausts contain benzene, a known leukemia risk factor). Also, the wire configuration exposure rating was carried out only in the homes, thereby ignoring potential exposures during the rest of the day, such as school, playground, travel on school buses, and so forth. Furthermore, there is some evidence that EMF may be a nonlinear "window of opportunity" risk factor, with only certain levels of exposure (neither too low nor too high) conferring potential disease risk.

An association between EMF and leukemia is *biologically plausible*. Although as nonionizing radiation EMF does not have enough energy to break DNA bonds, it has been theorized that increased EMF decreases synthesis of melatonin, a cancer-reducing agent. This hypothesis has received some support from animal and human experimental data.[7] However, the leukemias investigated in various EMF studies are not all of one specific type. Striking dissimilarities in the pathogenesis, pathophysiology, and prognosis of leukemia subtypes suggest that their etiologies may also differ greatly.

The *coherence* of the EMF-leukemia association measures its agreement with other known relationships with childhood leukemias and their natural history. Strong evidence supports an association between *in utero* exposure to ionizing radiation and childhood leukemia. By extension, prenatal exposure to other potential carcinogens might cause leukemia; an association with EMF exposure, then, is also possible.

Further coherence could be provided if an association was observed between EMF and birth defects (there are very few studies on the topic).

Considerable support or refutation of a causal association can be obtained from controlled laboratory *experiments*. A body of work on EMF effects on humans does exist, but the conditions under which those experiments are carried out (direct contact or near proximity between EMF source and target) are vastly different from environmental exposure, where the inverse-square decrease in EMF intensity with increasing distance is a primary consideration. In addition, little is known about assessing the eddies that are formed as EMF passes through the body.

If similar associations have proven to be causal, then by *analogy*, an EMF-leukemia relationship is more likely to be causal. But there are no clearly analogous causal associations for leukemia. Although the relationship between *in utero* ionizing radiation and leukemia mentioned earlier appears to be causal, nonionizing radiation such as EMF is quite different from ionizing radiation in nature and in its effects; there is, then, little analogy between these types of radiation.

In summary, the leukemia-EMF association certainly falls short of causality in strength, consistency, specificity, and analogy. Based on the depth of knowledge displayed in current literature, one cannot say that there is a causal association between EMF and leukemia. Conversely, there is insufficient information available to declare that the association is spurious or indirect.

The Bradford Hill postulates, or any other set of criteria for causality, can operate as a general outline for assessing epidemiologic associations. Every case to which such criteria are applied will "score" well on some points, poorly on others. Some of the criteria may be all but irrelevant to the issues surrounding the association. Still, examining how well such causal criteria can be applied to a conjectured association may serve to focus on appropriate directions for continued investigation. Searching, for example, for appropriate analogies or delineation of a dose-response curve may lead to important and useful evidence regarding the association, and may even lead to novel risk factor-disease associations to be evaluated.

REFERENCES

1. Dorn, H. F., Tobacco consumption and mortality from cancer and other diseases, *Public Health Rep.*, 74, 581, 1959.
2. Doll, R. and Hill, A. B., Mortality in relation to smoking: ten years' observation of British doctors, *Br. Med. J.*, 1:1399-1410; 1460-1467, 1964.
3. Evans, A. S., Causation and disease: the Henle-Koch postulates revisited, *Yale J. Biol. Med.*, 49, 175, 1976.
4. Surgeon General, Advisory Committee of the U.S. Public Health Service, Smoking and Health, PHS Pub. No. 1103, Public Health Service, Washington, D.C., 1964.
5. Hill, A. B., The environment and disease: association or causation?, *Proc. R. Soc. Med.*, 58, 295, 1965.
6. Wertheimer, N. and Leeper, E., Electrical wiring configurations and childhood cancer, *Am. J. Epidemiol.*, 109, 273, 1979.

7. Wilson, B.W. and Anderson, L. E., ELF electromagnetic field effects on the pineal gland in extremely low frequency electromagnetic fields, in *Extremely Low Frequency Electromagnetic Fields: The Question of Cancer*, Wilson, B. W., Stevens, R. G., and Anderson, L. E., Eds., Battelle Press, Columbus, OH, 159, 1990.

CHAPTER 4

Statistical Issues in the Design, Analysis, and Interpretation of Environmental Epidemiologic Studies

Gary M. Marsh

I. INTRODUCTION

A primary objective of many environmental epidemiological investigations is to associate potentially adverse exposures received at the workplace or in the community setting with potential biologic effects, thereby elucidating cause-effect relationships. Such associations are strengthened considerably if exposure-response relationships can be found, that is, if increasing levels of exposure are associated with increasing frequency of the biological effect.

In the workplace environment the identification and characterization of exposure-response relationships are facilitated by the general availability of personnel and industrial hygiene records. From these records investigators can enumerate the population at risk, determine the morbid or vital status of individual study members, characterize individual lifetime work histories and exposure profiles, and account for certain behavioral or lifestyle attributes, such as cigarette smoking or alcohol consumption, that can potentially confound or distort the interpretation of a true exposure-response relationship.

In contrast to occupational studies, the identification and characterization of exposure-response relationships in the residential community setting are made difficult by the general absence of documented data on individuals at risk and their potential exposures. Other problematic features of most community-based evaluations of health effects are the sheer diversity of situations in which potentially toxic substances and human exposures can be involved and the complex real-life situations, such as safeguarding confidentiality, that must be addressed.

More specifically, epidemiological studies of communities exposed to environmental contaminants are likely to be limited by the following technical and human problems:

0-87371-573-X/95/$0.00+$.50

- Populations living in the vicinity of a point source of exposure (e.g., a toxic waste site) are usually small, thus limiting both the range of outcomes and the size of the effects that can be studied.
- Persons living in any given area are usually heterogeneous either with respect to characteristics that can influence many health outcomes independently of exposure (e.g., age, race, socioeconomic status, occupation, smoking) or with respect to the type, level, duration, or timing of exposure. Also, there is in- and out-migration and geographical mobility within areas.
- Actual population exposures are generally poorly defined and for many chemicals or combinations of chemicals little or nothing is known about toxicological effects.
- Many of the health endpoints of interest are either rare (e.g., congenital malformations), are associated with long or variable latency periods (e.g., cancer), or are unlikely to have been recorded prior to the investigation (e.g., spontaneous abortions). In addition, the instruments used to measure health outcomes (e.g., questionnaires) are generally very insensitive.
- Publicity related to the episode under study may produce or accentuate reporting bias.
- The conduct of community studies is made difficult by the presence of a highly charged atmosphere of anger and fear, which often accompanies suspicion of adverse health effects.

Such limitations not only prohibit the development of a unified analytic approach to exposure and health outcome assessment but also prevent the generalization of statistical inferences drawn about a specific study population. How these numerous methodological limitations generally impact on the statistical aspects of community-based investigations of health effects is the topic of this chapter. The special problems associated with evaluating health effects of hazardous waste sites are reviewed by Marsh and Caplan.[1]

II. BASIC STATISTICAL CONCEPTS AND TERMINOLOGY

The discussion of the statistical issues in environmental studies requires an understanding of fundamental concepts and basic terminology, which are reviewed briefly in this section.

Target population: The group of individuals to whom one wishes to apply or extrapolate the results of an investigation. The target population may be, and often is, different from the population from which the sample in an investigation is drawn.

Random sample: A subset of a population that is used to draw conclusions or make estimates about attributes of the population. A probability sample is one in which all elements (persons) in the population have a known non-zero probability of being selected.

Variable: A characteristic for which measurements are made in a study. A quantitative variable is one that can be measured on a continuous scale (e.g., age, weight, height), whereas a qualitative variable conveys, through categorization or ranking, information regarding attribute (e.g., gender, race, illness severity). A random variable is one whose value arises as a result of chance factors.

Risk factor: A characteristic that has been shown to be associated with an increased probability of developing a disease or condition. A risk factor does not necessarily imply a cause and effect relationship.

Confidence interval: Two numerical values defining an interval that, with a specified level of confidence, is believed to include the population parameter being estimated.

Null hypothesis: The assertion that no true association or difference between variables exists in the population from which the study random samples are obtained.

Statistical significance test: A statistical procedure for determining the probability that the difference or association observed in a random sample might occur by chance factors alone if there is no true difference or association in the population (i.e., the null hypothesis is true). A two-tailed test is one in which the direction of the hypothesized difference (i.e., positive or negative) is not known, whereas a one-tailed test pre-establishes this direction.

P-value: Under the assumption that the null hypothesis is true, the probability of obtaining a difference or association as extreme or more extreme as the one actually observed.

Type I error: A statistical error that occurs when data from a random sample demonstrate a statistically significant result when no true difference or association exists in the population (i.e., the rejection of the null hypothesis when it is true). The significance (or alpha) level is the predetermined level of the type I error that will be tolerated, usually 5%. In a two-tailed statistical test the type I error is evenly divided between the hypothesized positive or negative differences.

Type II error: A statistical error that occurs when data from a random sample fail to demonstrate a statistically significant result when a true difference or association exists in the population (i.e., the failure to reject the null hypothesis when it is false). A general convention is to tolerate a type II (or beta) error of roughly no larger than four times the type I error, that is, about 20%.

Statistical power: The probability that an investigation will demonstrate a statistically significant result when a true difference or association of a specified level exists in the population (i.e., the rejection of the null hypothesis when it is false). For any given level of the true difference or association, power equals 1 minus the type II error. In most situations, one-tailed statistical significance tests are uniformly more powerful than their two-tailed counterparts.

Prevalence: The proportion of persons with a particular disease or condition at a point or period in time.

Incidence rate: The number of individuals who develop a disease or condition over a period of time divided by the number of person-years at risk.

Ecological study: A study in which the units of observation are groups of individuals rather than the individuals themselves.

Observational study: A nonexperimental study conducted at the individual level in which the circumstances of exposure cannot be controlled by the investigator. While most epidemiological investigations are observational, efforts can be directed at identifying naturally occurring exposure conditions that most closely simulate a controlled experiment.

Cohort study: An observational study in which two or more groups of persons who are free of disease and differ by extent of exposure to a potential cause of disease are compared over time with respect to the incidence of the disease. Also called follow-up, longitudinal, or prospective study.

Case-control study: An observational study in which a group of persons with a disease (cases) and a group of persons without the disease (controls) are identified without knowledge of prior exposure history and are compared with respect to exposure history. Case-control studies (or retrospective studies) can be used to estimate an hypothesized association that is derivable directly only in a cohort study. Controls can be selected as a simple random sample (unmatched) or matched to the cases on a group or individual basis. Matching is done to help control confounding bias and to increase the statistical efficiency of the comparisons.

Cross-sectional study: An observational study, usually a survey, which at the same point in time jointly classifies persons relative to disease status and exposure status. Cross-sectional studies are of limited value in estimating an hypothesized association between exposure and disease.

Relative risk: A measure of the strength of an hypothesized association derived from a cohort study found by taking the ratio of the incidence rate of disease among exposed persons to the same rate among the unexposed.

Odds ratio: A measure of the strength of an hypothesized association derived from a case-control study found by taking the ratio of the odds of exposure (the ratio of the proportion exposed to the proportion unexposed) among the cases to the same odds among the controls. When the disease incidence is low (say under 20%) the odds ratio is a good estimator of the relative risk.

Precision: Precision refers to the level of random or nonsystematic error associated with epidemiologic measurements. The primary component of random error is sampling error, which indicates the amount of variation in a measurement that would be obtained if similar studies were repeated a large number of times. Precision is reflected in the variance of a measurement and its associated confidence interval. Precision can be improved by increasing the size of a study or by modifying its design to increase the efficiency with which information is obtained from a given number of subjects.

Validity: Validity refers to the level of systematic error associated with epidemiological measurements. Systematic error, or bias, occurs if there is a difference between what the study is actually estimating and what it is intended to estimate. Systematic error, unlike random error, is not necessarily improved by increasing sample size. Validity is usually separated into two components: the internal validity, or degree to which the study findings truly represent the phenomena observed among the study sample, and external validity, or the extent to which the study findings can be generalized to persons outside the study population. A study must have internal validity before it can have external validity.

III. STATISTICAL ISSUES COMMON TO ALL STUDIES

A. EXPLORATORY VS. CONFIRMATORY STUDIES

Before a proper interpretation of data from an environmental epidemiology study can be made it is critical that the basic inferential nature of the study (or an analysis within a given study) be identified and understood. This nature can be simply classified into two broad groups of studies (or analyses), confirmatory and exploratory. An exploratory study (or analysis) is concerned with a completely new data set and there may be little or no prior knowledge about the problem. In contrast, a confirmatory study (or analysis) is designed primarily to test a hypothesis generated from a previous exploratory data analysis or speculated on theoretical grounds.

The related problem that arises is that statistical hypotheses are often generated and tested using the same data set. In principle, a statistical significance test should only be used to test a null hypothesis which is specified before observing the data, perhaps by using underlying theory or previous exploratory analyses. If an investigator selects the most unusual feature of a data set and then tests for its statistical significance using the same data, then the significance level of the test needs to be adjusted as he/she has effectively carried out multiple testing (see Sections III.B.4 and IV.E). It is generally desirable to confirm an observed effect on two or more data sets, not only to get a valid test but also to get results that generalize to different situations.

B. DETERMINING SAMPLE SIZE AND STATISTICAL POWER

In planning and interpreting environmental studies, sample size and power are extremely important considerations, as they help to determine study design and provide an objective basis from which study results can be meaningfully interpreted. The discussion below describes the relationship between statistical power, sample size, and a number of other important study parameters.

Specifically, statistical power per se is a function of the following study parameters:

1. *The size of the study and control groups.* In general, power increases as sample size increases.
2. *The background rate of the health outcome or exposure under study.* In cohort studies, power is directly related to the incidence rate of the disease among the nonexposed persons in the target population. In case-control studies, power is directly related to the relative frequency of exposure among the controls in the target population.
3. *The predetermined type I error rate.* With all other study parameters fixed, power

Table 1 Probability of Falsely Claiming Statistical Significance [1 – (1 α)ⁿ] at a Level α in at Least One of n Independent Comparisons

n	$\alpha = .05$ $1 - .95^n$	$\alpha = 0.01$ $1 - 0.99^n$
1	0.0500	0.0100
2	0.0975	0.0199
5	0.2262	0.0490
10	0.4013	0.0956
25	0.7226	0.2222
50	0.9231	0.3949
100	0.9941	0.6339

(or 1 – type II error) is directly related to the type I error rate. The type I and II errors are themselves inversely related.

4. *The number of associations being tested.* The probability of falsely concluding that an association is present at least once is affected by the number of associations that is tested within a single analysis. This phenomenon, referred to generally as the "multiple comparisons problem", arises frequently in epidemiologic research. Stated more precisely, the probability of falsely claiming statistical significance at a level α in at least one of n independent comparisons is $1 - (1 - \alpha)^n$. Table 1 shows this value, known as the experiment-wise error rate for selected values of α and n. This table reveals that the cumulative probability of making at least one type I error can be substantially larger when multiple hypotheses are tested. While the experiment-wise error rate can be controlled statistically via various simultaneous inferential procedures, the net effect is usually a decrease in overall power. Alternatively, the multiple comparisons problem is probably better controlled in environmental studies by limiting the number of exposure variables and/or health outcomes in the inferential analysis.
5. *The magnitude of the expected association between exposure and outcome.* With all other parameters fixed, power is directly related to this magnitude, which is often referred to as effect size. It is critical in designing an investigation that the investigator pre-establish the minimum effect size that it is important to detect statistically (on clinical, public health, or biological grounds only) at the acceptable type I and II error levels. These three study parameters (effect size, type I and II error) then determine the required sample size for the study.
6. *The design of the study and statistical techniques used for analysis.* There are several special design and analytic techniques that may be used to enhance power. These include: refining the history of exposure to avoid misclassification bias, refining the response variable to conform with an anticipated biologically coherent health outcome, increasing sample size via intensified case finding, forming composite exposure or outcome variables, use of continuous rather than discrete health outcome variables, use of repeated measures on each study member, stratification or matching, and clustering techniques.

The interrelationships of the primary study parameters that determine statistical power are illustrated in Figures 1 and 2. Figure 1, which pertains to cohort or cross-sectional studies, shows the relationships between the background incidence rate, sample size (in study and control group), and the magnitude of the effect (R) that can be demonstrated at the two-tailed 5% significance level with a power of 80%. For

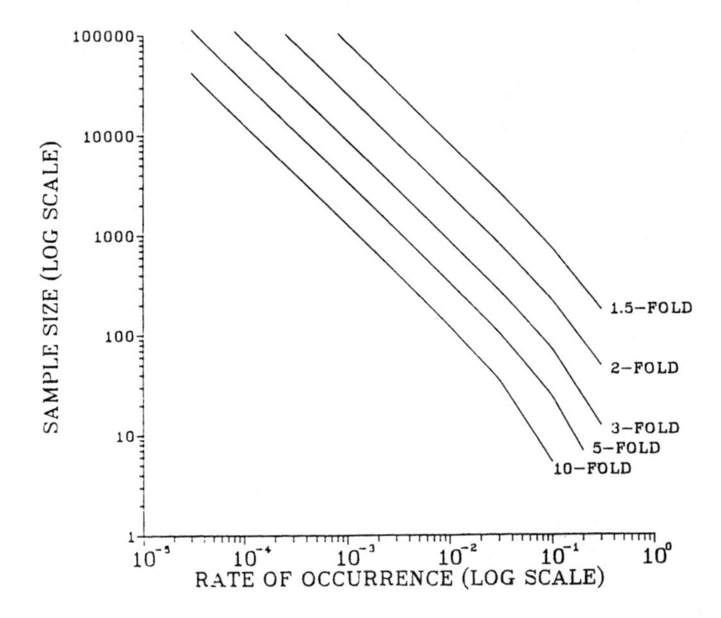

Figure 1 Cohort or cross-sectional study: relationship among sample size, background incidence rate, and effect size (R). $\alpha = 0.05$ (two-tailed), Power = 0.80

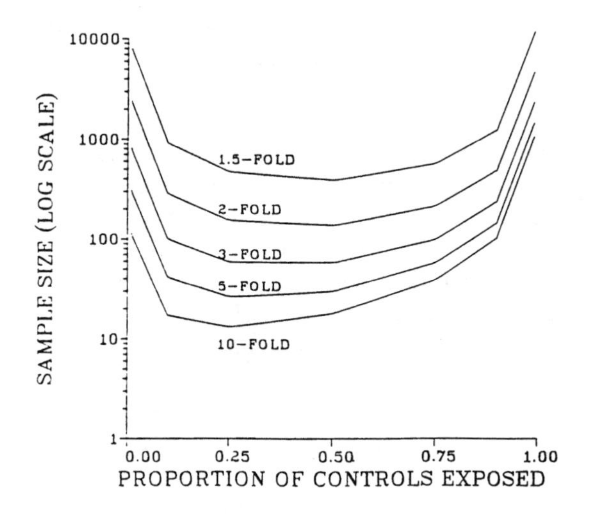

Figure 2 Unmatched case-control study: relationship among sample size, background exposure rate, and effect size (R). $\alpha = 0.05$ (two-tailed), Power = 0.80

example, to detect at these statistical error levels a twofold increase in illness prevalence from 0.1 to 0.2%, would require about 25,500 unexposed and 25,500 exposed persons, whereas a tenfold increase over the same background rate (0.1 to 1.0%) would require only about 1000 persons in each group.

Table 2 General Background Frequencies and Units of Analysis for Selected Health Outcomes

Health Outcome	Frequency	Unit of Analysis
Reproductive effects[a]		
Azoospermia	1×10^{-2}	Males
Birthweight <2500 g	7×10^{-2}	Live births
Spontaneous abortion after 8–28 weeks gestation	$1–2 \times 10^{-1}$	Pregnancies
Chromosomal anomaly among spontaneously aborted conceptions	$3–4 \times 10^{-1}$	Spontaneous abortions
Birth defects	$2–3 \times 10^{-2}$	Live births
Neural tube defects	$1 \times 10^{-4} – 1 \times 10^{-2}$	Live/still births
Cancer incidence[b]		
All sites	3.2×10^{-3}	Individuals
Stomach	9.8×10^{-5}	Individuals
Colon	3.3×10^{-4}	Individuals
Lung and bronchus	4.5×10^{-4}	Individuals
Bladder	1.5×10^{-4}	Individuals
Kidney	6.4×10^{-5}	Individuals
Lymphomas	1.2×10^{-4}	Individuals
Leukemias	9.5×10^{-5}	Individuals
Mortality[c]		
All causes	9.5×10^{-3}	Individuals
All cancer sites	1.6×10^{-3}	Individuals
Cirrhosis of liver	1.5×10^{-4}	Individuals
Congenital anomalies	8.4×10^{-5}	Individuals

[a] From Bloom A.D., Ed., *Guidelines for Studies of Human Populations Exposed to Mutagen and Reproductive Hazards*, March of Dimes Birth Defects Foundation, White Plains, NY, 1981.

[b] Average annual age-adjusted (1970) incidence rates, all S.E.E.R. sites.

[c] 1970 U.S. mortality rates.

In general, Figure 1 shows that for a given sample size, the power to discern modest effects increases with increasing frequency of the event under study, or for a given level of frequency the ability to detect a given effect size increases with increasing sample size. Table 2, which provides the frequencies of occurrence of selected health outcomes common to environmental studies, places the relationships shown in Figures 1 and 2 in a realistic perspective. The power in cohort studies can also be enhanced by increasing the control group size relative to the size of the study group.

In a similar fashion, Figure 2 shows for unmatched case-control studies the relationship between the proportion of controls exposed, sample size (in case and control group), and the minimum relative risk that can be detected at the two-tailed 5% significance level with a power of 80%. For example, in a community where 25% of the unaffected population is exposed to a potentially toxic material, an unmatched case-control study conducted at these statistical error levels would require about 150 cases and 150 controls to demonstrate a twofold increase in risk, whereas a tenfold increase in risk could be detected with only about 15 cases and 15 controls.

In general, Figure 2 shows that for a given case and control group size, the power to discern modest effects is maximized when the proportion of exposed controls is around 0.25 to 0.50. Conversely, for a given proportion of exposed controls, the ability to detect a given effect size increases with increasing sample size. The power in case-control studies can be enhanced by individual matching and/or by selecting multiple controls for each case.

C. IDENTIFYING AND CONTROLLING SYSTEMATIC SOURCES OF ERROR

Because the ultimate aim of most environmental studies is to describe an exposure-response relationship that is unlikely to be explained by extraneous differences between the two study groups, it is imperative that two sources of variation be controlled: variation in the characteristics of the study groups that relate to the *a priori* chance of exposure or to outcome and variation in the quality of data collected for the two study groups. The inability to control for these sources of variation can lead to a biased estimate of the exposure-outcome relationship.

These and other sources of bias resulting from methodological features of study design and analysis can be classified in a variety of ways, such as the catalog of biases provided by Sackett.[2] The following discussion describes the primary types of biases that can arise in environmental epidemiologic studies of community populations.

Confounding bias — Confounding bias, or confounding, arises from the failure to account for (or control for) the effects of other factors related to the exposure and health outcome. If no other biases are present three conditions are necessary for a variable to be a confounder: (1) a confounding variable must be a risk factor for the disease, (2) a confounding variable must be associated with the exposure under study in the population from which the cases derive, and (3) a confounding variable must not be an intermediate step in the causal pathway between exposure and disease. Confounding, which represents the most difficult bias to detect and correct, can be controlled in the study design (e.g., by matching in case-control studies), in the statistical analysis (e.g., by stratification), or both.

Selection bias — This term refers to a distortion in the estimate of effect resulting from the manner in which subjects are selected for the study population. Sources of selection bias include: flaws in the study design — most notably concerning the choice of groups to be compared (all types of studies), the choice of the sampling frame (particularly case-control and cross-sectional studies), loss to follow-up or nonresponse during data collection (in cohort studies), and selective survival (in case-control and cross-sectional studies). Selection bias can also result in case-control studies when the procedure used to identify disease status varies with exposure status.

Information bias — This bias, referred to as "misclassification", is related to the instruments and techniques used to collect information on exposure, health outcomes, or other study factors. Nondifferential information bias occurs when the likelihood of misclassification is the same for both groups being compared. When an effect exists, bias from nondifferential misclassification is always in the direction of the association stated in the null hypothesis (i.e., no effect), and thus is of particular concern in studies that show no association between exposure and disease. Differential information bias occurs when the likelihood of misclassification is different for each comparison group. This form of bias is potentially more problematic because it can bias the observed effect estimate either toward or away from the null value.

Reporting bias — There are certain difficulties associated with investigating an aware and exposed population living in the vicinity of a source of environmental contamination that extend beyond the methodologic issues that are statistically or epidemiologically tractable. For example, when community reports of adverse events are the impetus of subsequent health studies, statistical concerns arise that the hypothesis has been suggested by the data (see Section III.A). Usual tests of significance may be inappropriate as one has chosen to test for an effect that has already been noted. While the observation, ideally, should be tested in another area, this may not be possible due to the uniqueness of the exposure or to other reasons that require the investigation to be confirmed. When faced with this dilemma, approaches are needed that compensate for the positive bias without introducing negative bias. Excluding a reported cluster can negatively bias an exposure-outcome relationship if the community is small, the outcome is rare, and most of the current cases have been found. A cohort study on a truly independent sample could be mounted if the exposures were continuing; however, this design is generally not feasible for studying rare health outcomes.

D. ASSESSING INTERACTION

Persons involved in most environmental studies are exposed via many different routes to mixtures or combinations of several potentially toxic chemicals. When more than one exposure is involved, not only must the existence of distinct relationships between each exposure and health outcome be verified, but also the relationships among the exposures must be examined. For example, one exposure may confound another, leading to confounding bias in the estimate of the exposure-outcome relationship. In addition, the first exposure may interact with the second by potentiating or suppressing the relationship between the second exposure and health outcome. In this case the relationship of the second exposure to the outcome changes with the level of the first exposure.

Several authors have considered the statistical/epidemiologic issue of interaction as it applies to the combined effect of two or more exposures.[3–6] Although analytic methods have been developed to assess interactions, their application to environmental studies may be limited because: the most useful and interpretable analysis of interaction requires the application of multivariate statistical techniques and the typically weak and incomplete data derived from environmental studies may not be amenable to these more sophisticated modes of analysis, and several types of statistical models are available for assessing interaction but there is some dispute over which is the most appropriate.[4–7] The relevance of these statistical models to the biology of environmentally related illness will remain uncertain, however, until more is known about how various chemical exposures produce illness or biologic effects.

E. DETERMINING THE LEVEL OF THE INVESTIGATION

Environmental health effects investigations can be classified into three levels.[8] Level I is based on existing, routine, and easily accessible exposure and health outcome records. For example, level I studies may draw on vital certificate data or special registries of tumors or malformations in order to examine birthweights, perinatal mortality, cancer incidence or mortality, or sex of offspring. Level I studies will lack power since they will usually be limited to poorly defined measures of exposure and estimates of exposure-outcome relationships may not be adjustable for the effects of potentially confounding factors. Level I investigations include the large class of ecologic studies.

Level II includes short-term epidemiologic studies, such as cross-sectional, case-control, or short-term cohort, that require the collection of more precise, individual exposure and health outcome data, as well as data on potentially confounding variables. The decision to proceed from the level I descriptive evaluation to level II analytic studies is based on the following criteria.[9]

1. A statistically significant association is observed at level I but there is a need to explore a broader spectrum of outcomes and/or to specify more precisely the exposure-outcome relationship (e.g., threshold and dose-response components of the exposure).
2. The endpoint that stimulated the investigation, or that seems most biologically coherent with the exposure, is not accessible to level I studies.
3. The results in level I are inconclusive.

In level II studies, the statistical considerations of power, bias, and interaction are applicable to the choice of study design, enabling the researcher to make maximal use of small numbers, rare events, and uncertain information sources. Level II studies can entertain a wide range of endpoints and can include outcomes identified in medical records (spontaneous abortions, malformations, behavioral or psychological disorders), through interviews with study subjects (spontaneous abortion, sexual dysfunction, symptoms or signs of rashes, paralysis, eye irritation), or through biological studies of study subjects (biochemical, immunologic, and chromosomal assessments, and nerve conduction velocities).

Level III involves well-planned, long-term investigations, such as cohort studies of exposed and unexposed residential cohorts. Since this design is well suited for diseases with long latency periods it has been considered mainly for the purpose of discovering environmental carcinogens. Level III studies are greatly facilitated by the existence of centralized and accessible registries of births, deaths, and diseases.

IV. OTHER MORE SPECIALIZED STATISTICAL ISSUES

A. PROBLEMS WITH ECOLOGICAL STUDIES

Because ecologic studies are based on existing data on large populations, they are well suited as a preliminary or exploratory approach to evaluating health effects of environmental exposures. Despite this practical advantage, causal inference about individual events from grouped data is limited by the following methodological problems:[11]

1. Data on many extraneous risk factors (age, sex, occupation, personal habits) may not be available on the ecologic level, thus preventing control over confounding bias on the observed exposure-health outcome relationships. For example, geographic variations in mortality are probably more likely due to differences in socioeconomic status, alcohol and tobacco consumption, and diet rather than to any common environmental exposure, unless that exposure is very intense or exceptionally toxic.
2. The populations at risk very often leave the study prior to the investigation so that examining cross-sectional data with ecologic analysis can lead to serious selection bias.
3. A particular limitation with ecologic time trend analysis is that generally only mortality data are available. Morbidity data, when available, are difficult to interpret due to either changes in ascertainment or changes in definitions of disease. Moreover, trend data can be greatly affected by extraneous factors, such as the prevalence of chronic disease in the population (an indicator of susceptibility to effects of toxic exposures), and also more powerful factors, such as weather, respiratory infections, and natural disasters.
4. In ecologic analysis, there is the potential for substantial bias in effect modification. This problem, known as the ecological fallacy, occurs when the composition of each group is not homogeneous with respect to the study factor. Theoretically, the bias resulting from ecologic analysis can make an association appear stronger or weaker than it is at the individual level; however, in practice, this bias ordinarily exaggerates the magnitude of a true association, if one exists.
5. With ecologic analysis, certain predictor variables (especially sociodemographic and environmental variables) tend to be more highly correlated with each other than they are at the individual level — a phenomenon called "multicolinearity". Consequently, the increased correlations between predictor variables make it particularly difficult to isolate their effects on the health outcome. In general, multicolinearity is most problematic for ecologic studies involving larger and fewer geographically defined units of analysis.

These and other problems with ecologic analysis, such as measurement error and ambiguity of cause and effect, can be minimized, however, via the following approaches:[11] (1) use of ecologic regression rather than correlation, including in the model as many risk factors as possible, (2) use of data that are grouped into the smallest geographic units of analyses as possible, subject to the constraints of intergroup migration and unstable rate estimation, and (3) attempt to ascertain how the groups were formed and analyze using all variables thought to be related to the grouping process.

B. ADDITIONAL SAMPLE SIZE AND POWER CONSIDERATIONS

There are a number of other less well-known factors that impact on the statistical power or sample size characteristics of environmental epidemiology studies. For example, McKeown-Eyssen and Thomas[12] examined the influence of the distribution of exposure on sample size calculations and showed that power calculations based on the dichotomization of continuous exposure variables will generally overestimate the actual sample size needed to detect a given effect size of interest. Their results are particularly relevant to exposure distributions with extremely high or low variability or which are highly skewed (such as radon exposures, which are usually lognormally distributed in populations at risk of exposure).

Also, Smith and Day[13] have quantified the extent to which the confounding effects of study covariables influence sample size requirements in case-control studies. Based on their approach, it would be necessary, for example, to increase the sample size 20 to 60% to detect a twofold increase in risk (i.e., effect size of 2) from the primary exposure variable after adjusting for confounders in which there is a strong exposure-confounder relationship.

Finally, the methods of Smith and Day,[13] Thomas and Greenland,[14] and Greenland[15] can be used to determine the power for testing interaction effects. These methods have shown that if the aim of a study is to detect important interactions, the size of the study will have to be at least four times larger than if attention were confined to detecting main effect sizes of the same magnitude.

C. LIMITATIONS OF MULTIVARIATE MODELS

In environmental studies involving a large number of exposure and potentially confounding variables conventional descriptive or stratified methods of statistical analysis may not be feasible, especially if the study variables are time-related or there is an interest in examining interactions among the exposure and/or confounding variables. In relatively recent years, considerable advances have been made in the development of multivariate mathematical models and related computer software for analyzing such complex data. One class of models that is particularly well-suited for the analysis of prevalence or incidence data is the multiplicative model. Commonly used multiplicative models include logistic regression, Poisson regression, and the Cox proportional hazards model. If used appropriately, these models provide a useful framework for testing hypotheses, adjusting for potential confounders, increasing statistical power, and exploring biologic mechanisms. These and other models are discussed in detail in Breslow and Day[16,17] and Checkoway et al.[18]

The proper application of these models for complex data also requires that a more complex set of mathematical assumptions be satisfied, and analyses based on these models can be quite sensitive to deviations from these assumptions. For example, the logistic model involves a number of strong assumptions, notably the exponentiality of the exposure-response for continuous variables and the multiplicativeness of interactions between multiple variables. If such assumptions are violated alternative parametric models should be considered using, for example, a method for fitting "general relative risk models"[19] by constructing nonparametric exposure-response relations and testing goodness-of-fit for continuous variables.[20]

D. CONFOUNDER IDENTIFICATION

In multivariate modeling of exposure-response relationships particular attention must be placed on "overadjusting" for potentially confounding effects. Overadjustment can arise by: (1) including a variable that is intermediate on a causal pathway from an exposure to a disease, (2) inclusion of a variable highly correlated with the exposure but not causally related to disease (in a case-control study, for example, this would inflate the variance of the odds ratio but would not introduce bias), and (3) including only those variables that reduce the estimated odds ratio for the association of interest.[21] Determining whether a variable is a true confounder requires epidemiologic judgement and consideration of biologic credibility and cannot be simply resolved by statistical analysis of the apparent associations in the data.[22]

E. THE PROBLEM OF MULTIPLE INFERENCE

Because most environmental studies involve multiple hypotheses, the risk of type I error needs to be addressed. There are two aspects to this problem, multiple biologic hypotheses and multiple ways of characterizing each hypothesis in terms of statistical significance tests. One approach to minimizing the multiple inference problem with respect to the latter aspect is to construct from one or more individual variables a single composite variable for each exposure that is believed, on *a priori* grounds, to predict risk most strongly. These constructed variables can then be used in multivariate models rather than the individual variables. Secondary analyses of potential effect modifiers (variables that modify or interact with the effects of other variables) are then used to search for stronger and simpler summary measures of exposure. This approach avoids the excessive data dredging inherent in multivariate analyses with many alternative indices of the same risk factor.

Regarding the problem of multiple biological hypotheses, there exist a number of formal statistical procedures for adjusting for multiple comparisons, such as multiplying the P-value by the number of statistical tests, that are appropriate. This and other approaches are described by Thomas et al.[23]

F. THE UTILITY OF META-ANALYSIS

The term meta-analysis (literally, the analysis of analyses) is commonly used to denote the complete range of quantitative methods developed to facilitate the synthesis of results from different research studies of a related hypothesis. Such a synthesis is a central problem when attempting to review and interpret a body of knowledge published in the scientific literature. Typically, this body of knowledge is based on a number of studies that differ with respect to design, analysis, validity, reliability, and conclusions.

Although a vast array of specific meta-analysis methods exist, most involve a quantitative summary of results across studies investigating a common research question. In effect, all meta-analysis methods treat the study per se as the individual unit of statistical analysis. Recent applications of meta-analysis in environmental epidemiology include the effects of passive smoking,[24] the relationship between asbestos exposure and gastrointestinal cancer,[25] and the relationship between water chlorination by-products and certain cancers.[26]

The advantages of meta-analysis include the potential for increased precision in risk estimates, a more complete and systematic literature review than would otherwise be performed, and a more formal assessment of the consistency of findings across various studies. The key disadvantages of meta-analysis in environmental epidemiology include the frequent omission of unpublished "negative" studies (the so-called file drawer problem), the inability to evaluate the quality of the individual studies considered in a review, difficulties in combining studies that differ in design, method for controlling for potential confounding, or approach to exposure assessment. A comprehensive review and critique of meta-analysis methods is provided by Greenland.[27]

REFERENCES

1. Marsh, G. M. and Caplan, R. J., Evaluating health effects of exposure at hazardous waste sites: a review of the state-of-the-art, with recommendations for future research, in *Health Effects from Hazardous Waste Sites*, Andelman, J. B. and Underhill, D. W., Eds., Lewis Publishers, Chelsea, Michigan, 1987, chap. 1.
2. Sackett, D. L., Bias in analytic research, *J. Chron. Dis.*, 32, 51, 1979.
3. Koopman, J. S., Causal models and sources of interaction, *Am. J. Epidemiol.*, 106, 439, 1977.
4. Kupper, L. L. and Hogan, M. D., Interaction in epidemiologic studies, *Am. J. Epidemiol.*, 108, 447, 1978.
5. Rothman, K. J., Synergy and antagonism in cause effect relationships, *Am. J. Epidemiol.*, 99, 385, 1974.
6. Rothman, K. J., Occam's razor pares the choice among statistical models, *Am. J. Epidemiol.*, 108, 347, 1978.
7. Walter, S. C. and Holford, T. R., Additive, multiplicative, and other models for disease risks, *Am. J. Epidemiol.*, 208, 314, 1978.
8. Report of Panel II, Guidelines for reproductive studies in exposed human populations, in *Guidelines for Studies of Human Populations Exposed to Mutagen and Reproductive Hazards*, Bloom, A. D., Ed., March of Dimes Birth Defects Foundation, New York, 1981.
9. Hill, A. B., The environment and disease: association or causation, *Proc. R. Soc. Med.*, 58, 295, 1965.
10. Monson, R. R., *Occupational Epidemiology*, 2nd ed., CRC Press, Boca Raton, 1990.
11. Morgenstern, H., Use of ecologic analysis in epidemiologic research, *Am. J. Public Health*, 72, 1336, 1983.

12. McKeown-Eyssen, G. and Thomas, D. C., Sample size determination for case-control studies: the influence of the distribution of exposure, *J. Chron. Dis.*, 38, 559, 1985.
13. Smith, P. and Day, N. E., The design of case-control studies: the influence of confounding and interaction effects, *Int. J. Epidemiol.*, 13, 356, 1984.
14. Thomas, D. C. and Greenland, S., The efficiency of matching in case-control studies of risk-factor interactions, *J. Chron. Dis.*, 38, 569, 1985.
15. Greenland, S., Tests for interaction in epidemiologic studies: a comparative review and a study of power, *Stat. Med.*, 2, 243, 1983.
16. Breslow, N. E. and Day, N. E., The analysis of case-control studies, in *Statistical Methods in Cancer Research*, International Agency for Research on Cancer, Lyon, 1980, 1.
17. Breslow, N. E. and Day, N. E., The design and analysis of cohort studies, in *Statistical Methods in Cancer Research*, International Agency for Research on Cancer, Lyon, 1987, 2.
18. Checkoway, H., Pearce, N. E., and Crawford-Brown, D. J., *Research Methods in Occupational Epidemiology*, Oxford University Press, Oxford, 1989.
19. Thomas, D. C., General relative-risk models for survival time and matched case-control analysis, *Biometrics*, 37, 673, 1981.
20. Thomas, D. C., Nonparametric estimation and tests of fit for dose-response relations, *Biometrics*, 39, 263, 1983.
21. Day, N. E., Byar, D. P., and Green, S. B., Overadjustment in case-control studies, *Am. J. Epidemiol.*, 112, 696, 1980.
22. Robins, J. M. and Greenland, S., The role of model selection in causal inference from nonexperimental data, *Am. J. Epidemiol.*, 123, 392, 1986.
23. Thomas, D., Siemiatycki, J., Dewar, R., Robins, J., Goldberg, M., and Armstrong, B. G., The problem of multiple inference in studies designed to generate hypotheses, *Am. J. Epidemiol.*, 122, 1080, 1985.
24. Environmental Tobacco Smoke: Measuring Exposures and Assessing Health Effects, National Academy Press, Washington, D.C., 1986.
25. Frumkin, H. and Berlin, J., Asbestos exposure and gastrointestinal malignancy review and meta analysis, *Am. J. Ind. Med.*, 14, 79, 1988.
26. Morris, R. D., Audet, A. M., Angelillo, I. F., Chalmers, T. C., and Mosteller, F., Chlorination, chlorination by-products, and cancer: a meta-analysis, *Am. J. Public Health*, 82, 955, 1992.
27. Greenland, S., Quantitative methods in the review of epidemiologic literature, *Epidemiol. Rev.*, 9, 1, 1987.

CHAPTER 5

Biological Markers of Exposure

Marilyn F. Vine and Barbara S. Hulka

The purpose of this chapter is to define what biological markers of exposure are, present examples of biological markers of exposure, discuss the criteria upon which a researcher would decide to use a biological marker of exposure in an epidemiologic investigation, and provide an example of how these criteria could be applied to select a biological marker of exposure for a specific study question.

I. WHAT ARE BIOLOGICAL MARKERS OF EXPOSURE?

Biological markers of exposure are a subset of biological markers in general. Biological markers can be defined as "cellular, biochemical, or molecular alterations that are measurable in biological media such as human tissues, cells or fluids" (Hulka, 1990). As shown in Figure 1, a modified version of a figure presented in a report on *Biologic Markers in Reproductive Toxicology*,[41] biological markers can be classified into categories that represent a sequence of events from exposure to disease, including markers of internal dose, biologically effective dose, early response, altered structure and function, and disease. Outside of this sequence of events are susceptibility factors that can influence events at any point along the pathway. This classification system is based on work presented by Perera and Weinstein.[34] The sequence of events from exposure to disease is initiated when an individual is exposed to a chemical substance in the external environment and is thought to progress unless the agent is eliminated from the body or resultant damage along the way is repaired.

For the purposes of this discussion, a biological marker of exposure is considered to be "a qualitative or quantitative marker of internal dose or biologically effective dose".[48] More specifically, a marker of internal dose indicates the presence of an exogenous chemical, either unchanged or metabolically altered, in the body of an individual. Table 1 presents examples of markers of exposure that can be classified as markers of internal dose.

0-87371-573-X/95/$0.00+$.50

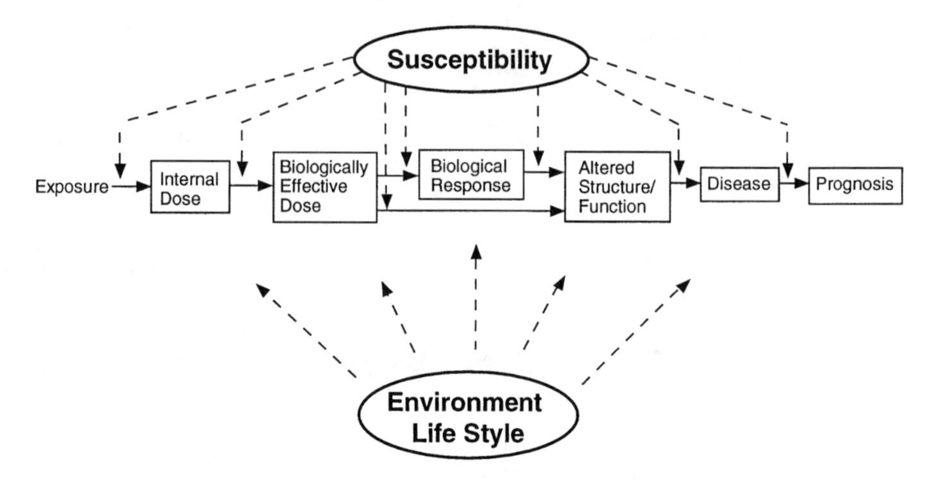

Figure 1 The relationship of biological markers to exposure and disease. (Modified from *Biologic Markers in Reproductive Toxicology*, National Academy Press, 1989. With permission.)

Table 1 Examples of Markers of Internal Dose

Internal Dose Marker	Exposure	Biological Media	Unchanged/ Metabolically Altered
Lead	Gasoline, paint, lead smelter emissions	Blood, hair, nails, bone, teeth	Unchanged
Selenium	Food	Urine, feces, serum, nails, hair	Unchanged
Fluoride	Water, toothpaste, mouth rinses	Blood, urine, hair, nails, saliva, bone	Unchanged
Cotinine	Tobacco products	Blood, urine, saliva, cervical mucus, semen, breast milk	Metabolic byproduct of nicotine
Mutagenesis assays	Tobacco products, chemotherapy agents, other mutagens	Urine, feces	Often metabollically activated compounds

A marker of biologically effective dose indicates the amount of absorbed chemical that has "interacted with critical subcellular targets, measured either in the target tissue or a surrogate tissue".[25] Examples of markers of exposure that would fall into the category of biologically effective dose markers include DNA and protein adducts (Table 2). DNA adducts are chemicals covalently bonded with DNA. An example of a DNA adduct is pictured in Figure 2, benzo(*a*)pyrenediol-epoxide adducted to the DNA base, guanine. The presence of DNA adducts in various tissues, such as lung, bladder, placenta, or white blood cells may result, for example, from exposures to cigarette smoke[11,14,35] or coke oven emissions.[22] Protein adducts consist of chemicals covalently bonded to proteins such as ethylene oxide, used, for example, in the sterilization of hospital equipment, covalently bonded to the amino acid, histidine, in

Table 2 Markers of Biologically Effective Dose

Biologically Effective Dose Markers	Exposure	Biological Media
DNA adducts	Tobacco smoke, coke oven emissions	Tissues (lungs, heart), white blood cells
Protein adducts	Tobacco smoke, ethylene oxide	Red blood cells, albumin

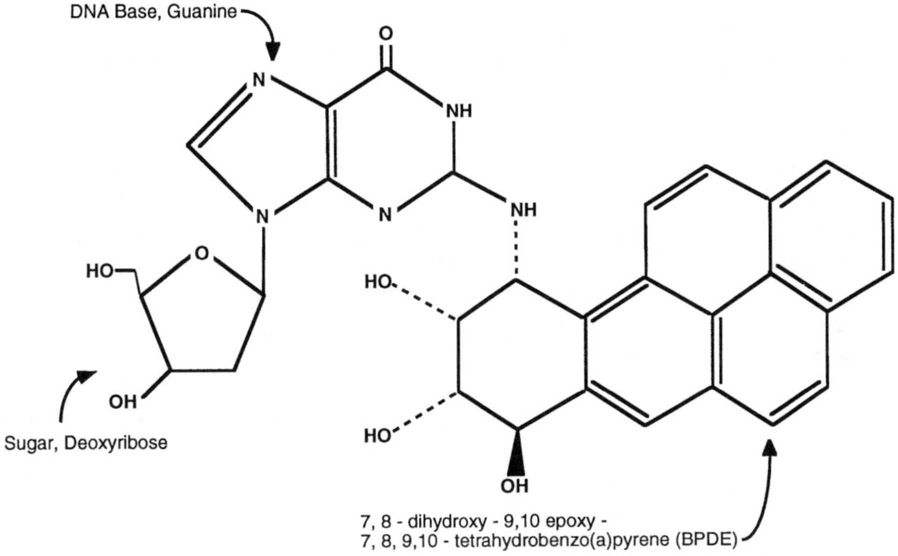

Figure 2 Benzo(*a*)pyrene DNA adduct: example of a DNA adduct formed through the covalent bonding of the DNA base guanine with benzo(*a*)pyrene after metabolic activation to 7,8-dihyroxy-9,10-epoxy-7,8,9,10- tetrahydrobenzo(*a*)pyrene (BPDE). (From Stowers SJ, Anderson MW: *Environ. Health Perspect.* 62:31-40 (1985). Reprinted with permission.)

the hemoglobin of red blood cells (Figure 3). For a more in depth discussion of biological markers in general, see *Biological Markers in Epidemiology*.[26]

II. WHY USE BIOLOGICAL MARKERS OF EXPOSURE IN ENVIRONMENTAL EPIDEMIOLOGIC INVESTIGATIONS?

The most difficult aspect of environmental epidemiologic research is accurately characterizing an individual's level of exposure. Traditionally, epidemiologists have characterized exposure to environmental agents according to the following categorization schemes: (1) whether the individual has ever been exposed to the agent of interest (yes/no), (2) the duration of exposure, (3) the degree of exposure relative to

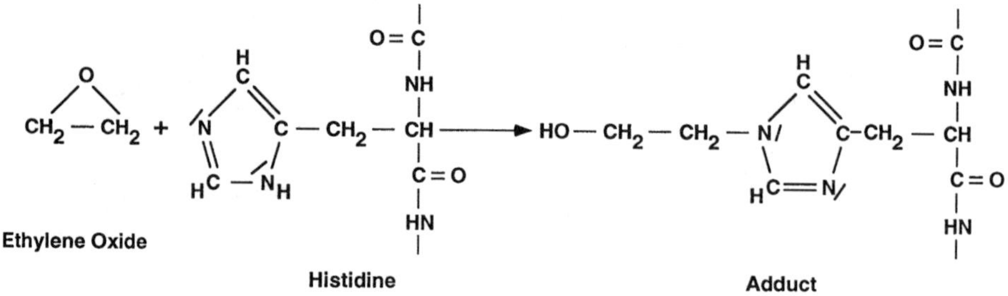

Ethylene Oxide

Histidine

Adduct

Figure 3 Ethylene oxide hemoglobin adduct. Ethylene oxide reacts *in vivo* with the imidazole nitrogens of the histidyl residues of hemoglobin. (From Calleman CJ, et al.: *J. Environ. Pathol. Toxicol. Oncol.* 2:427-442 (1978). Reprinted with permission.)

the exposure of other individuals (high, medium, low), (4) the level of exposure in the environment immediately surrounding the individual, obtained, for example, through air monitoring, and (5) quantified personal measurements such as those that can be obtained with a radiation badge worn by the individual.[7] Quantified personal measurements are the most desirable measures of exposure status. Biological markers of exposure are being evaluated to clarify their advantages in providing better estimates of exposure than those relying on measurements external to the human body.

The expectation is that biological markers of exposure used in conjunction with, or independently of, traditional exposure measures will help improve current estimates of exposure in environmental epidemiologic research. The hope, then, is that the use of biological markers will allow researchers to more accurately assess exposures to individuals, thereby improving the sensitivity of epidemiologic studies to detect associations (especially weak associations) between exposures and disease outcomes. This can be accomplished by reducing misclassification of exposure measures and by providing better quantitative assessments of exposures which can be used to determine the extent of dose-response relationships between exposures and disease outcomes.

Biological markers of exposure have the potential to provide better estimates of the exposures that are relevant to possible health effects than other measures of exposure. For example, while area air monitoring will provide information as to levels of air pollutants in the environment, internal dose markers can provide information concerning the amount of the exogenous agent that actually entered the body, taking into account the use of protective clothing, personal hygiene practices, etc. Biologically effective dose markers go one step further and indicate that the agent reached a particular target tissue of interest.

Markers of exposure may be used to validate the use of other measures of exposure. For example, during the late 1980s there was concern over the potential health effects of eating polychlorinated biphenyl (PCB)-contaminated fish from Lake Michigan. In one study, conducted by Dar et al.,[12] assessing the association between fish consumption and adverse reproductive effects among women in Green Bay, Wisconsin, serum PCB levels were compared to estimates of fish consumption determined by questionnaires among a subgroup of the women in the study. It was determined that fish consumption was a reasonably good estimate of serum PCB levels (Pearson correlation of 0.666), and, therefore, could be used to assess PCB exposure. This eliminated the need to analyze blood specimens for PCB levels among all of the women, a costly and time-consuming procedure.

In addition, the use of exposure markers may help (1) to determine the extent of human exposure to exogenous agents through population monitoring of markers in tissues such as blood and fat, (2) to improve knowledge of compliance in intervention trials by assessing the presence of markers of the intervention in the blood or urine, for example, of the study participants, and (3) to improve knowledge of mechanisms of disease by providing clues to intermediate steps in the pathway from exposure to disease.[26,48] While not every marker will be able to fulfill every goal, the properties of markers of exposure as a group that enable them to potentially fulfill the above mentioned goals are listed below.

A. MARKERS OF EXPOSURE PROVIDE INFORMATION SPECIFIC TO THE INDIVIDUAL

Often in environmental and occupational studies researchers have to rely on ecologic measures of exposure such as area air monitoring or job titles that assign the same exposure to a group of individuals. The exposure measure in many studies of the health effects of fluoride exposure, for example, has been derived from estimates of water fluoridation in the county or city water supplies. This measure does not take into consideration the amount of water consumed by each individual nor exposure that has resulted from other sources of fluoride, such as bottled water, toothpaste, and mouth rinses. By taking samples of blood, for example, one may have a better chance of determining the degree of fluoride exposure to each individual.

B. MARKERS OF EXPOSURE ARE ABLE TO INTEGRATE EXPOSURES OVER TIME AND FROM VARIOUS ROUTES OF ENTRY INTO THE BODY

Exposure assessment can be difficult when the level of the exposure fluctuates over time. Exposure measurements made intermittently, whether through traditional means or the use of biological markers, may not accurately reflect an individual's overall exposure. In fact, more general questionnaire information or area monitoring may provide a more accurate and consistent estimate of exposure to the individual than fluctuating levels of a marker measured sporadically. Markers that integrate the dose over time may improve the precision of the exposure measurement.[48]

Further complicating exposure assessment is the fact that the same chemical contaminant can be found in a variety of media, including air, water, soil, food, etc. As a result, an exogenous chemical may enter the body through different routes, such as inhalation, ingestion, or absorption through the skin. Measuring the level of the contaminant in only one environmental medium might underestimate exposure. Measuring levels of the contaminant in all media would be expensive and time consuming. A marker that would integrate the exposure dose over time and from various routes of entry into the body could give a more accurate picture of the overall exposure to the individual. Hemoglobin adducts, measured in red blood cells, are examples of such a marker. Hemoglobin adducts accumulate during the life of the red blood cell, which is about 120 d.[39] The half-life of adipose tissue is about 1.5 years among individuals who are not losing or gaining weight.[43] Therefore, adipose tissue integrates exposures from various routes of entry into the body over a long period of time. For example, the commercial supply of milk on the island of Oahu was found to be contaminated with heptachlor epoxide during 1981 to 1982 as the result of contaminated cattle feed.[2] Heptachlor is a pesticide that accumulates in adipose tissue. Researchers could have sampled the adipose tissue of potentially exposed individuals to assess the extent of heptachlor epoxide exposure. However, sampling adipose tissue is a fairly invasive procedure. Because a reliable equilibrium ratio exists between adipose tissue and serum levels,[1] blood samples could potentially have been used to obtain an estimate of the long-term integrated exposure to heptachlor.

C. MARKERS OF EXPOSURE OFTEN ALLOW THE QUANTIFICATION OF EXPOSURE DOSE

Markers differ in their ability to distinguish the amount of exposure to the individual. Some markers can only distinguish the presence or absence of exposure but not level of exposure. Some markers can provide information on the relative, but not absolute, levels of exposure. DNA adducts as detected by ^{32}P-postlabeling fall into this category.[15] Other markers provide a more quantitative estimate of exposure dose, some only within specific ranges of dose. 4-aminobiphenyl hemoglobin adducts, for example, show a linear dose-response relationship with the number of cigarettes smoked per day under 20 cigarettes per day. Above 20 cigarettes per day the curve levels off.[39]

D. MARKERS OF EXPOSURE CAN IDENTIFY SPECIFIC AGENTS WITHIN A MIXTURE

Markers differ in their ability to distinguish the specific environmental agent to which the individual was exposed. Some markers, such as urine mutagenesis assays, can provide information as to whether or not the individual was exposed to a mutagenic substance, but not necessarily which specific agent.[36] Mutagenesis assays have often been used in workplace situations to assess whether something in the work environment was mutagenic and not whether the workers were exposed to a particular mutagenic agent. Other assays, such as the enzyme-linked immunosorbent assay,[16] will detect specific mutagenic agents within a complex mixture. If a person was exposed to a complex mixture, such as tobacco smoke, the ELISA method could detect the presence of a specific agent in the mixture, for example benzo(*a*)pyrene, by assessing the presence of specific DNA adducts, in this case benzo(*a*)pryenediol-epoxide (BPDE) DNA adducts, in the cells of the exposed individuals.[37]

E. MARKERS OF EXPOSURE CAN RELATE TIME OF EXPOSURE TO DOSE

Hair and nails are examples of tissues that store chemical contaminants providing a record of an individual's exposure history as the tissues grow. Heavy metals, such as lead for example, bind to sulfhydryl groups in the hair. The amount of heavy metals present in the hair closest to the scalp would indicate more recent exposures while heavy metals in more distal hair would indicate past exposures.[48] Knowing the rate of growth of the hair would provide information on the time of exposure. An analogous situation occurs with chemicals such as selenium that are incorporated into finger and toenails.

Properties of biological markers briefly described above are presented in greater detail in Wilcosky and Wilcosky and Griffith.[47,48]

III. CRITERIA FOR SELECTING A BIOLOGICAL MARKER OF EXPOSURE

The first step in determining whether or not to use a biological marker of exposure in an environmental epidemiologic investigation is to clearly define the objectives of the study. These objectives will serve as the focal points for all future decisions. Once the objectives are clearly defined, the decision to use a biological marker of exposure and the selection of which marker to use must be based on both biologically and laboratory-related criteria. There are many issues to be considered.

Biologically Related Criteria:

1. What are the properties (physical, chemical, and/or biological) of the external exposure to be measured? Can the exposure be measured unchanged or is it metabolically altered in the body?
2. What markers are available to be used to measure the exposure of interest?
3. How specific is the marker for the particular exposure?
4. In which biological media can the marker be measured?
5. How invasive is the technique for obtaining biological specimens?
6. How soon can the marker be measured in the body after the exposure has occurred?
7. How long does the marker persist in the body?
8. Is the internal exposure dose a measure of the peak dose or a measure of the dose integrated over time?
9. What is the variability of the marker in the population, both in terms of intraindividual variability and interindividual variability?
10. How stable is the marker under various storage conditions?

Laboratory-Related Criteria:

1. How available is the assay?
2. What is the cost of the assay?
3. What is the sensitivity of the assay in terms of the minimum levels of the exposure it can detect?

4. Can the marker be quantified in such a way as to show dose-response relationships between the external exposure and the level of the marker?
5. What is the specificity of the assay? What exactly does it measure?
6. How reliable is the assay?

In this section, we will illustrate how to select a marker of exposure for an epidemiologic investigation by considering the criteria for marker selection in relation to a specific research question.

What are the objectives of the study? Suppose that one wanted to study the effect of maternal exposure to environmental tobacco smoke (ETS) on adverse reproductive outcomes, such as low birthweight of the offspring. Active and passive cigarette smoke exposure have been reported to be associated with low birthweight[19,31] and low birthweight is a risk factor for morbidity and mortality among infants.[9] As a study objective, one may want to determine the levels of maternal ETS exposure that are associated with low birthweight. Participants in the study would include non-smoking women who are exposed to ETS to varying degrees (including no exposure) during the nine months of pregnancy.

A. BIOLOGICALLY RELATED CRITERIA

1. What are the Properties (Physical, Chemical, and/or Biological) of the Exposure to be Measured? Can the Exposure be Measured Unchanged or is it Metabolically Altered in the Body?

ETS consists mostly of sidestream smoke, the smoke that is released from the burning end of the cigarette when the smoker is not puffing. The remainder of the ETS consists of mainstream smoke, the smoke that is exhaled by the smoker, and the smoke that is emitted from the cigarette as the smoker puffs.[8] ETS is an aerosol that is composed of thousands of chemicals that often undergo secondary chemical reactions. Currently, there is no way to directly measure the amount of ETS absorbed by an individual. Exposure to ETS can be assessed by the administration of questionnaires, by measuring the concentrations of certain ETS constituents in the air, and by measuring levels of biological markers.[8] Some chemical constituents of ETS may enter the body unchanged, such as cadmium, whereas others are metabolically altered. Nicotine, for example, can be measured in its original form or as its metabolite, cotinine.

2. What Markers are Available to be Used to Measure the Exposure of Interest?

Because ETS contains many chemical constituents, many markers are potentially useful in epidemiologic studies. Possible markers include: levels of nicotine, its metabolic byproduct, cotinine; thiocyanate (a metabolic byproduct of hydrogen cyanide present in tobacco smoke); exhaled carbon monoxide; carboxyhemoglobin; urine mutagenicity; DNA adducts; and hemoglobin adducts.

3. How Specific is the Marker for the Particular Exposure?

In order to be sure that the marker of exposure is measuring ETS exposure among the women in the study and not exposure to other combustion processes or other chemical exposures, it is helpful if the marker selected is as specific as possible to ETS. Nicotine and cotinine occur in body fluids virtually exclusively as the result of exposure to tobacco products.[42] Nicotine chewing gum and other smoking cessation aids contain nicotine. Some foods and teas also contain nicotine, but one would have to consume large amounts to significantly affect cotinine levels.[13] Hemoglobin adducts associated with exposure to the nitrosamines, 4-methylnitrosamino-1-(3-pyridyl)-1-butanone (NNK) and N′-nitrosonornicotine (NNN), are also tobacco-specific.[23] Hemoglobin adducts of 4-ABP may also serve as markers of ETS exposure since the main environmental source of 4-ABP exposure is tobacco smoke.[13] Nonspecific markers of tobacco smoke exposure include thiocyanate, which can also be found in certain leafy vegetables and nuts,[20,29] exhaled carbon monoxide and levels of carboxyhemoglobin that can result from other combustion processes, such as the burning of gasoline in cars, and urine mutagenicity, which can occur as the result of exposure to a variety of mutagenic agents.

ETS and mainstream smoke contain many of the same chemicals. For the current study, women who are active smokers should be excluded. If for some reason active smokers are not excluded, it is usually possible to distinguish active from passive smokers by determining levels of various markers. For example, the mean level of cotinine in the urine of nonsmokers is less than 1% of the mean cotinine level among active smokers.[8,44]

The question still remains as to whether the particular marker of tobacco smoke chosen for the study represents the agent(s) within ETS that affects the outcome of interest, in this case low birthweight. One would have to know the mechanism by which exposure to ETS might lead to low birthweight in order to determine the answer to this question. It should be noted, however, that a marker does not have to measure a causal agent in order to be useful in predicting an adverse outcome. The marker could be correlated with the causal agent.

4. In Which Biological Media Can the Marker be Measured?

From the above discussion of specificity, one can see that nicotine, cotinine and hemoglobin adducts associated with exposure to NNN, NNK, and 4-ABP are among the most specific markers of tobacco smoke exposure. Nicotine and cotinine can be measured in about 1 ml of a variety of body fluids, including urine, saliva, blood, breast milk, and cervical mucus. Hemoglobin adducts can be measured in about 10 ml of whole blood.

5. How Invasive is the Technique For Obtaining Biological Specimens?

The degree of invasiveness of the technique necessary to obtain the biological specimens will influence the numbers and characteristics of the participants in the study — the lower the degree of invasiveness, the greater the likelihood that people will participate in the study. Most of the fluids mentioned above can be obtained in a fairly noninvasive manner. Venipuncture to obtain blood is the most invasive technique, yet it is a procedure that is routinely done.

6. How Soon can the Marker be Measured in the Body After the Exposure Has Occurred?

The issue of time to appearance of the marker is influenced by the absorption, metabolism, and distribution of the exogenous agent and is more important with respect to acute (one-time exposures) and intermittent exposures than with continuous exposures. If measurements are made too soon, one may falsely conclude that the person was not exposed. Cotinine, for example, can be measured in body fluids shortly (within minutes to hours) after exposure to ETS.[10] Hemoglobin adducts should be detectable in a relatively short period of time after exposure as well.

7. How Long does the Marker Persist in the Body?

Persistence of the marker also bears on the issue of when to assess the presence of the marker. Persistence is influenced by the metabolism, storage, and excretion of the exogenous agent and also by the stability of the biological medium in which the marker is measured. Again, this is more of a concern for acute or intermittent exposures than with continuous exposures. If one waits until the marker has disappeared, for example, because the kidneys excreted it out of the body, then one would falsely conclude that the person had not been exposed. Nicotine has a very short half-life, only about 2 h,[4] making intermittent exposures to ETS difficult to detect. The half-life of cotinine in various fluids among active smokers is approximately 15 to 16 h.[21,32] Because of this length of persistence, cotinine is more suited to assessing daily exposure. Among nonsmokers the half-life has been reported to be twice as long.[21,38] Thus, if cotinine were the marker of choice, one could detect ETS exposures to the women in the study that occurred a day or two prior to the sampling of the biological specimens used to assess cotinine levels. Hemoglobin adducts, on the other hand, are thought to persist for the life of the red blood cell, about 120 d[39] and would, therefore, integrate exposures over a longer period of time.

8. Is the Internal Dose a Measure of the Peak Dose or a Measure of the Dose Integrated Over Time?

Whether the dose is peak or integrated depends on whether the exposure is a one-time phenomenon or a continuous exposure, and whether or not the exogenous agent is stored in the tissue sampled. If the agent is not stored in the tissue sampled, then typically what happens is that for one-time exposures, levels of the marker will increase for a certain period after the exposure (peak) and then decrease as the body excretes, metabolizes, or stores the foreign agent in another body tissue. If the exposure is chronic, then a steady state can be achieved between the amount of chemical (marker) entering and leaving the tissue. For example, since the blood does not store cotinine, with chronic ETS exposure plasma cotinine levels would represent a measure of the integrated dose over time. If one assessed a tissue that stored the agent of interest, one could get an assessment of the accumulated dose.

9. What is the Variability of the Marker in the Population?

Intraindividual variability refers to differences in the value of the marker over time and within different biological media within the same individual. Intraindividual variability in marker levels is influenced by daily and seasonal rhythms, the tissue sampled, and changes in the exposure over time. The cotinine concentration in the urine, for example, is an order of magnitude higher than the cotinine concentration in the blood of the same individual.[44] Differences in cotinine concentrations are likely to result if one assesses the cotinine concentration in the urine of a nonsmoker in the morning vs. the end of a work day spent in the company of smokers.[28]

Interindividual variability, or the differences in marker levels between individuals, is influenced by differences in metabolism, genetic make-up, and, for some markers, DNA repair capabilities. Cotinine metabolism, for example, varies by smoking status[21,38] and race.[46] Very little is known about the interindividual variability of hemoglobin adducts. However, 4-ABP hemoglobin adduct levels have been found to vary with acetylation status and oxidation phenotype among both active and passive smokers.[3] Thus, 4-ABP hemoglobin adduct levels are not strictly a measure of the level of tobacco smoke exposure, but may indicate the biologically effective dose of 4-ABP with respect to certain conditions such as bladder cancer.

10. How Stable is the Marker Under Various Storage Conditions?

The stability of the marker will influence whether the samples must be analyzed immediately or whether they can be stored and analyzed later. These considerations will influence the feasibility of the study. Cotinine, for example, is very stable over time when stored at –20°C or below.[24] The stability of hemoglobin adducts of NNN and NNK over long periods of time is under investigation. Decisions concerning how samples are stored may dictate future uses of the samples.

B. LABORATORY-RELATED CRITERIA

1. How Available is the Assay?

The availability of the assay to measure the marker of interest depends on the complexity of the assay and the demand for the assay. Some assays are commercially available, such that an epidemiologist could send the samples to a laboratory and have them analyzed for a fee. However, there are several assays that are in the development phase or are so complicated and time consuming that only one or two laboratories in the world perform them. Cotinine has become the preferred marker of tobacco smoke exposure. It can be detected either by radioimmunoassay (RIA) or with gas chromatography and mass spectrometry (GC-MS). The methods to detect cotinine are not very complicated and many laboratories are capable of performing the assays. However, some laboratories have better quality control procedures than others. Many of the hemoglobin adduct assays, on the other hand, are still in development. In some cases, only one laboratory in the country performs a particular hemoglobin adduct assay.

2. What is the Cost of the Assay?

Assay costs vary dramatically. The cotinine assay can be performed for under $20 per sample, whereas hemoglobin adduct assays can exceed $400 per sample. Cost is largely related to the labor intensiveness of the procedure. If the high cost of an assay does not cause the marker to be eliminated from consideration, it may reduce the sample size of the study.

3. What is the Sensitivity of the Assay?

Because ETS exposure is a fairly low-level exposure, especially in comparison with active smoke exposure, it is important that the assay technique be able to detect small quantities of the marker of interest. Radioimmunoassays are very sensitive, for example, detecting levels of cotinine as low as 0.5 ng/ml,[8] sufficient to detect ETS exposures. Similarly, assays used to detect 4-ABP hemoglobin adducts have sufficient sensitivity to detect ETS exposures (Maclure et al., 1989). No work has been done with regard to ETS and hemoglobin adducts of NNN and NNK.

4. Can the Marker be Quantified in Such a Way as to Reflect Dose-Response Relationships Between External Exposure and Level of the Marker?

As mentioned earlier, some markers are better at detecting dose-response relationships than others. Cotinine has been shown to indicate degrees of ETS exposure when measured in urine, plasma, saliva, and semen.[29,44] For example, cotinine concentrations in infants' urine increases with the number of cigarettes smoked by the mother in the previous 24 h.[18] Because the urine concentrates cotinine, and levels of cotinine in urine are higher than in other body fluids,[29,44] urine is considered the best fluid for detecting ETS exposure.[45]

Levels of 4-ABP hemoglobin adducts are linearly associated with active tobacco smoke exposure to 20 cigarettes smoked per day. After that the curve levels off.[39] 4-ABP hemoglobin adduct levels have been found to be associated with amount of ETS exposure, defined as high vs. low exposure based on questionnaire information and confirmed by cotinine levels.[33] As mentioned earlier, they can also vary with certain metabolic phenotypes.

Hemoglobin adducts associated with exposure to NNN and NNK may be quanitified by assessing the levels of 4-hydroxy-1-(3-pyridyl)-1-butanone (HPB) that are released upon treatment of the adducted hemoglobin with a base. However, levels of HPB are not correlated with numbers of cigarettes smoked per day or cotinine levels and only 20% of active smokers have elevated levels of HPB.[6] Therefore, NNN and NNK hemoglobin adducts are poor markers of the level of tobacco smoke exposure.

5. What is the Specificity of the Assay?

Different techniques can be used to measure a particular marker, but not all techniques measure the exact same thing. Nicotine and cotinine can be assessed with both RIA and GC-MS. RIAs sometimes detect agents that are similar to the agent of interest as well as the agent itself. With respect to the cotinine assay, there is little cross-reactivity with other metabolites of nicotine.[30] Both RIA and GC-MS techniques are susceptible to contamination of reagents and equipment.[8]

6. How Reliable is the Assay?

A reliable assay can be performed on the same sample many times and provide the same answer repeatedly. An unreliable assay can be a source of misclassification of the exposure measure in an environmental epidemiologic investigation. The coefficient of variation (standard deviation/mean) is a measure of the reliability of an assay. The lower the coefficient of variation the higher the reliability. Results of cotinine analyses with both RIA and GC-MS are highly repeatable,[8,17] indicating that these techniques are reliable. The intra- and interassay coefficient of variation for the RIA has been reported to be about 5%.[8]

IV. CONCLUSION REGARDING THE "BEST" MARKER OF ETS EXPOSURE

For the study involving ETS exposure of pregnant women and adverse reproductive outcomes, cotinine would probably be the most useful marker of exposure. Cotinine is essentially tobacco-specific and can be measured in a noninvasive manner in a small amount of various body fluids, including urine, saliva, and blood, with urine being the most desirable fluid for detecting ETS exposure. Measurement of cotinine in body fluids indicates exposure to ETS over the previous day or two. Periodic measurements during the 9 months of pregnancy can provide a good indication of ETS exposure. Cotinine is stable when frozen so that assays can be completed after all samples are collected. The assays for the detection of cotinine are inexpensive, reliable, relatively specific and sensitive, quantifying levels of exposure as low as 0.5 ng/ml. In a study of the effects of active smoking by the mother on low birthweight of the infant, Haddow et al.[19] found that mothers' serum cotinine levels, measured at 15 to 20 weeks gestation, were a better predictor of low birthweight than the number of cigarettes smoked per day by the mother.

V. CONCLUDING REMARKS

Biological markers are by no means the ultimate cure for all the ills of exposure assessment. It is hoped, however, that the use of markers of exposure, alone or together with other more traditional exposure information, will lead to improvements in exposure assessment in environmental epidemiologic investigations. The increased accuracy in the measurement of exposures that is expected with the use of exposure markers has the potential to enable environmental epidemiologic investigations to detect associations between exposures and health outcomes that may have been missed with other more traditional measures of exposure.

The properties of biological markers of exposure that may enable them to improve upon more traditional exposure measures are (1) they provide information specific to the individual, (2) they are able to integrate exposures over time and from various routes of entry into the body, (3) they often allow the quantification of exposure dose, not only to the individual, but to target or surrogate tissues of interest, (4) they can identify specific chemicals within a mixture, and (5) some markers can relate time of exposure to the dose received. Markers, however, should not be used indiscriminately. When deciding whether or not to employ a marker of exposure in an environmental epidemiologic study, careful thought must be given to the specific properties of the available marker(s) and how the marker will enable the investigator to answer the study question.

REFERENCES

1. Anderson HA: Utilization of adipose tissue biopsy in characterizing human halogenated hydrocarbon exposure. *Environ. Health Perspectives* 60:127-131 (1985).
2. Baker DB, Loo S, Barker J: Evaluation of human exposure to the heptachlor epoxide contamination of milk in Hawaii. *Hawaii Med. J.* 50(3):108-118 (1991).
3. Bartsch H, Caporaso N, Coda M, Kadlubar F, Malaveille, Skipper PL, Talaska G, Tannenbaum SR, Vineis P: Carcinogen hemoglobin adducts, urinary mutagenicity, and metabolic phenotype in active and passive cigarette smokers. *J. Natl. Cancer Inst.* 82:1826-1831 (1990).
4. Benowitz NL, Kuyt F, Jacob P III: Circadian blood nicotine concentrations during cigarette smoking. *Clin. Pharmacol. Ther.* 32:758-764 (1982).
5. Calleman CJ, Ehrenberg L, Jansson B, Osterman-Golkar S, Segerback D, Svensson K, Wachmeister CA: Monitoring and risk assessment by means of alkyl groups in hemoglobin in persons occupationally exposed to ethylene oxide. *J. Environ. Pathol. Toxicol. Oncol.* 2:427-442 (1978).
6. Carmella SG, Kaga SS, Kagan M, Foiles PG, Palladino G, Quart AM, Quart E, Hecht SS: Mass spectrometric analysis of tobacco-specific nitrosamine hemoglobin adducts in snuff dippers, smokers and nonsmokers. *Can. Res.* 50:5438-5445 (1990).
7. Checkoway H, Pearce NE, Crawford-Brown DJ. *Research Methods in Occupational Epidemiology.* New York: Oxford University Press, p.22 (1989).
8. Committee on Passive Smoking: *Environmental Tobacco Smoke — Measuring Exposures and Assessing Health Effects.* Washington, D.C.: National Academy Press (1986).

9. Committee to Study the Prevention of Low Birthweight, Division of Health Promotion and Disease Prevention, Institute of Medicine: *Preventing Low Birthweight.* Washington, D.C.: National Academy Press, 1985.
10. Curvall M, Enzell CR: Monitoring absorption by means of determination of nicotine and cotinine. *Arch. Toxicol. Suppl.* 9:88-102 (1986).
11. Cuzick J, Routledge MN, Jenkins D, Garner RC: DNA adducts in different tissues of smokers and nonsmokers. *Int. J. Can.* 45:673-678 (1990).
12. Dar E, Kanarek MS, Anderson A, Sonzogni WC: Fish consumption and reproductive outcomes in Green Bay, Wisconsin. *Environ. Res.* 59:189-210 (1992).
13. Environmental Protection Agency. *Respiratory Health Effects of Passive Smoking: Lung Cancer and Other Disorders* pp. 3-42 (EPA 600/6-90/006F). Washington, D.C.: U.S. Environmental Protection Agency (1992).
14. Everson RB, Randerath E, Santella RM, Cefalo RC, Avitts TA, Randerath K: Detection of smoking-related covalent DNA adducts in human placenta. *Science* 231:54-57 (1986).
15. Everson RB, Randerath E, Santella RM, Avitts TA, Weinstein IB, Randerath K: Quantitative associations between DNA damage in human placenta and maternal smoking and birthweight. *Science* 80(8):567-576 (1988).
16. Goldring JM, Lucier GW: Protein and DNA adducts. In: *Biological Markers in Epidemiology.* pp. 78-104. Hulka BS, Wilcosky TC, Griffith JD (eds). New York: Oxford University Press (1990).
17. Gori GB, Lynch CJ: Analytical cigarette yields as predictors of smoke bioavailability. *Regul. Toxicol. Pharmacol.* 5:314-326 (1985).
18. Greenberg RA, Haley NJ, Etzel RA, Loda FA: Measuring the exposure of infants to tobacco smoke. *N. Engl. J. Med.* 310:1075-1078 (1984).
19. Haddow JE, Knight GJ, Palomaki GE, Kloza EM, Wald NJ: Cigarette consumption and serum cotinine in relation to birthweight. *Br. J. Obstet. Gynecol.* 94:678-681 (1987).
20. Haley NJ, Axelrod CM, Tilton KA: Validation of self-reported smoking behavior: biochemical analyses of cotinine and thiocyanate. *Am. J. Public Health* 73(10): 1204-1207 (1983).
21. Haley NJ, Sepkovic DW, Hoffmann D: Elimination of cotinine from body fluids: disposition in smokers and nonsmokers. *Am. J. Public Health* 79(8):1046-1047 (1989).
22. Harris CC, Vahakangas K, Newman MJ, Trivers GE, Shamsuddin A, Sinopoli N, Mann DL, Wright WE: Detection of benzo(a)pyrene diol epoxide-DNA adducts in peripheral blood lymphocytes and antibodies to the adducts in serum from coke oven workers. *Proc. Natl. Acad. Sci. U.S.A.* 82:6672-6676 (1985).
23. Hecht SS, Hoffmann D: Tobacco-specific nitrosamines, an important group of carcinogens in tobacco and tobacco smoke. *Carcinogenesis* 9(6):875-884 (1988).
24. Hu, Ping-Chuan, personal communication (February, 1992).
25. Hulka BS, Wilcosky T: Biological markers in epidemiologic research. *Arch. Environ. Health* 43(2):83-89 (1988).
26. Hulka BS: Overview of biological markers. In: *Biological Markers in Epidemiology.* pp. 3-15. Hulka BS, Wilcosky TC, Griffith JD (eds). New York: Oxford University Press (1990).
27. Hulka BS, Wilcosky TC, Griffith JD (eds). *Biological Markers in Epidemiology.* New York: Oxford University Press (1990).
28. Jarvis MJ, Russell MAH: Measurement and estimation of smoke dosage to nonsmokers from environmental tobacco smoke. *Eur. J. Respir. Dis. Suppl.* 133:68-75 (1984).

29. Jarvis MJ: Application of biochemical intake markers to passive smoking measurement and risk estimation. *Mutat. Res.* 222:101-110 (1989).
30. Langone JJ, Gjika H, Van Vunakis H: Nicotine and its metabolites. Radioimmunoassays for nicotine and cotinine. *Biochemistry* 12:5025-5030 (1973).
31. Lazzaroni F, Bonassi S, Manniello E, Morcaldi L, Repetto E, Ruocco A, Calci A, Cortellessa G: Effect of passive smoking during pregnancy on selected perinatal parameters. *Int. J. Epidemiol.* 19(4):960-966 (1990).
32. Lynch CJ: Dose-measurements in humans. *Eur. J. Respir. Dis. Suppl.* 133:63-67 (1984).
33. Maclure M, Katz RBA, Bryant MS, Skipper PL, Tannenbaum SR: Elevated blood levels of carcinogens in passive smokers. *Am. J. Public Health* 79(10):1381-1384 (1989).
34. Perera FP, Weinstein IB: Molecular epidemiology and carcinogen-DNA adduct detection: new approaches to studies of human cancer causation. *J. Chron. Dis.* 35:581-600 (1982).
35. Perera FP, Santella RM, Brenner D, Poirier MC, Munshi AA, Fischman HK, Ryzin: DNA adducts, protein adducts, and sister chromatis exchange in cigarette smokers and nonsmokers. *J. Natl. Cancer Inst.* 79(3):449-456 (1987).
36. Rynard S: Urine mutagenicity assays. In: *Biological Markers in Epidemiology*. pp. 56-77. Hulka BS, Wilcosky TC, Griffith JD (eds). New York: Oxford University Press (1990).
37. Santella RM, Li Y, Zhang YT, Young TL, Stefanidis M, Lu XQ, Lee BM, Gomes M, Perera FP: Immunological methods for the detection of polycyclic aromatic hydrocarbon-DNA and protein adducts. In: *Genetic Toxicology of Complex Mixtures*. pp. 291-301. Waters MD et al. (eds). New York: Plenum Press, (1990).
38. Sepkovic DW, Haley NJ, Hoffmann D: Elimination from the body of tobacco products by smokers and passive smokers (letter). *JAMA* 256:863 (1986).
39. Skipper PL, Tannenbaum SR: Protein adducts in the molecular dosimetry of chemical carcinogens. *Carcinogenesis* 11(4):507-518 (1990).
40. Stowers SJ, Anderson MW: Formation and persistence of benzo(a)pyrene metabolite-DNA adducts. *Environ. Health Perspect.* 62:31-40 (1985).
41. Subcommittee on Reproductive and Neurodevelopmental Toxicology: *Biologic Markers in Reproductive Toxicology*. p. 17. Washington, D.C.: National Academy Press (1989).
42. Tunstall-Pedoe H, Woodward M, Brown CA: The drinking, passive smoking, smoking deception and serum cotinine in the Scottish Heart Health Study. *J. Clin. Epidemiol.* 44(12):1411-1414 (1991).
43. van Staveren WA, Deurenberg P, Katan MB, Burema J, de Groot L, Hoffmans M: Validity of the fatty acid composition of subcutaneous fat tissue microbiopsies as an estimate of the long-term average fatty acid composition of the diet of separate individuals. *Am. J. Epidemiol.* 123(3):455-463 (1986).
44. Vine MF, Hulka BS, Margolin BH, Truong YK, Hu PC, Schramm MM, Griffith JD, McCann M, Everson RB: Cotinine levels in smokers and nonsmokers. *Am. J. Public Health*, in press.
45. Wall MA, Johnson J, Jacob P, Benowitz NL: Cotinine in the serum, saliva and urine of nonsmokers, passive smokers, and active smokers. *Am. J. Public Health* 78(6): 699-701 (1988).

46. Wegenknecht LE, Cutter GR, Haley NJ, Sidney S, Manolio TA, Hughes GH, Jacobs DR: Racial differences in serum cotinine levels among smokers in the Coronary Artery Risk Development in (Young) Adults Study. *Am. J. Public Health* 80(9):1053-1056 (1990).
47. Wilcosky TC: Criteria for selecting and evaluating markers. In: *Biological Markers in Epidemiology*. pp. 28-55. Hulka BS, Wilcosky TC, Griffith JD (eds). New York: Oxford University Press (1990).
48. Wilcosky TC, Griffith JD: Applications of biological markers. In: *Biological Markers in Epidemiology*. pp.16-27. Hulka BS, Wilcosky TC, Griffith JD (eds). New York: Oxford University Press (1990).

CHAPTER 6

Epidemiologic Aspects of Environmental Hazards to Reproduction

Lowell E. Sever

I. INTRODUCTION

Love Canal, Woburn, Brownsville, St. Gabriel — these are names in the news. What do they have in common? All are communities where questions have been raised about possible associations between adverse reproductive outcomes and exposure to suspected hazardous substances in the environment.

The potential reproductive and developmental effects of exposure to environmental contaminants have become of increasing concern during the last two decades. With the recognition of effects of exogenous substances — drugs and chemicals — on human reproduction, attention began to focus on possible adverse reproductive outcomes associated with exposure to substances in the ambient environment. Attention has been directed to the possible reproductive effects of toxic waste sites, industrial facilities, the use of pesticides, and chemical contamination of food and our environment. Concerns have been expressed by citizens, scientists, regulators, and policy makers as knowledge about reproductive hazards has increased. Yet, our understanding of the magnitude of the risks, if any, associated with environmental toxins is far from complete. At a time when an estimated 70,000 chemicals are in everyday use and 500 to 1000 new ones are synthesized each year,[1] it is reasonable to ask what this change in our environment means to the health of potential parents and their unborn children.

Basic information about hazards needs to be obtained. There is growing interest in identifying the environmental factors that may contribute to reproductive and developmental toxicology, quantifying the risks involved, and then moving toward reduction in risks through regulation and other mechanisms of risk management. Public health personnel must be able to determine the adverse reproductive outcomes potentially associated with an exposure situation and design epidemiologic studies in which possible associations between exposure and outcome can be examined.

0-87371-573-X/95/$0.00+$.50

Epidemiologic methods and studies can play a central role in this process for prevention. In this chapter, some of the features of epidemiologic studies are discussed that set out to determine if there is an association between an environmental exposure and an adverse reproductive outcome.

The embryo/fetus is particularly sensitive to effects of environmental agents. While the data suggesting a major role of the ambient environment in abnormal development are far from compelling, there are reasons for concern. Substances such as pesticides, heavy metals, and organic solvents have been shown to have developmental effects in animals.[2] While human data are more limited, the example of developmental defects associated with exposure to organic mercury discussed later serves as a reminder of the potentially devastating effects of environmental contamination.[3] Some studies of chemical production facilities, metal smelters, environmental contamination, and toxic waste sites have suggested adverse reproductive effects associated with exposure of pregnant women.[3]

The selection and definition of the outcomes to be studied are important issues in designing an epidemiologic study of environmental reproductive hazards. Several authors have pointed out that a spectrum of adverse outcomes, ranging from reduced fertility to developmental disabilities and malignancies developing later in life, is potentially associated with exposure to environmental agents. The adverse outcomes observed have included spontaneous abortions, congenital malformations, low birthweight, developmental disabilities, and infant mortality.[4]

An important general issue is the biological relationship among adverse reproductive outcomes. It is necessary to look at an array of reproductive outcomes when attempting to determine if an environmental exposure has an adverse effect. Many endpoints may not be independent events; more than one endpoint may be related to the same exposure. There may be a spectrum of adverse outcomes that vary in frequency, severity, and type in relation to exposure timing and dose.

This chapter examines some of the issues, epidemiologic methods, and findings that suggest associations between environmental exposures and adverse reproductive outcomes. Some basic aspects of human reproduction and development are considered and some of the health endpoints that may be associated with exposure to environmental agents are discussed. Some aspects of the epidemiologic study of selected health endpoints are reviewed and some of the problems inherent in their study are discussed. We close by examining some specific examples of selected environmental reproductive hazards.

II. REPRODUCTION AND DEVELOPMENT

Every stage of the multistep process of reproduction can be disrupted by external environmental agents. Reproduction is a couple-specific (dependent) process and dealing with reproductive and developmental toxicity must address the couple-specific aspect of reproduction and the generation of a third individual during development. These events occur intermittently; the goal is a healthy child who is capable of perpetuating the process.

Reproductive effects are those that affect males and females prior to coitus and conception. These may include alterations in the production and functional maturation of germ cells (eggs and sperm), hormonal status, libido, and other factors that are essential both for normal reproductive function and the genetic integrity of the offspring. Thus, in reproductive effects we include factors that affect male or female fecundity and factors that lead to heritable genetic abnormalities, mutations at either the gene or chromosomal level.

There have been more studies of reproductive toxicity and reproductive effects of occupational exposures than of environmental exposures. As part of the growing attention to occupational reproductive hazards, increasing attention has been paid to the effects of chemical exposures on male reproductive function.[5] Studies have addressed the effects of a number of chemicals on sperm production and a pesticide, dibromochloropropane, has been established as a prototypic male reproductive toxicant.[5] Both direct and indirect approaches to studying male reproductive toxicity have been developed. Direct approaches include studying a number of aspects of semen, such as sperm count, shape, and motility.[6] Indirect approaches include studying the numbers of offspring in workers exposed to a substance of concern and comparing the observed number with an expected number, based on a standard population. This has been referred to as the standardized fertility ratio and has been promoted as less invasive than semen analysis for examining reproductive toxicity in males.[7]

Reproductive toxicity in females has been studied through evaluation of menstrual and endocrinological patterns.[8] In addition, an indirect method, time to pregnancy, has been developed for the assessment of reproductive toxicity in females.[9]

Interest in the effects of occupational and environmental exposures on reproductive function is growing. The reader is referred to a recent report by the National Research Council that addresses this topic[8] and to guidelines regarding male and female reproductive toxicity promulgated by the Environmental Protection Agency.[10,11]

Concern has developed regarding possible associations between childhood cancers and parental, particularly paternal, preconception exposures to hazardous agents. Attention has been focused on occupational exposures, including exposure to chemicals such as hydrocarbons,[12] ionizing radiation,[13] and electromagnetic fields.[14] Extensive discussion of this topic is beyond the scope of this chapter, but the interested reader is referred to reviews that explore this issue more fully.[15-17]

Developmental effects are those that occur after conception and which lead to structural or functional deviations in the conceptus. Developmental effects are due to direct impacts on the zygote, blastocyst, embryo, or fetus and can be manifest by death, structural abnormalities, interference with growth, or abnormal function. This process of induction of abnormal development is known as teratogenesis. The rest of this chapter will focus on teratogenesis and developmental effects.

III. TERATOGENESIS

Insufficient attention has been paid to the principles of teratogenesis in epidemiologic studies of adverse reproductive outcomes associated with environmental exposures. These principles should be emphasized in considering the outcomes that should be examined in such studies.

A teratogen is a substance, organism, or physical agent capable of causing abnormal development. Traditionally, the identification and definition of teratogenic agents were based on an agent's ability to produce structural defects. More recently, the concept of teratogenesis has been expanded to include agents that act during embryonic or fetal development to produce deviations from normal structure, function, or both. With this expanded view, the outcomes of teratogenic exposure are considered to include not only structural defects but functional abnormalities, growth retardation, and death of the organism.[3] Others use the term "developmental toxicity" to refer to this broad array of effects.

Wilson[18] has developed six general principles of teratogenesis that are relevant to the design and conduct of epidemiologic studies. These principles are

- The final manifestations of abnormal development are death, malformation, growth retardation, and functional disorder
- Susceptibility of the conceptus to teratogenic agents varies with the developmental stage at the time of exposure
- Teratogenic agents act in specific ways (mechanisms) on developing cells and tissues in initiating abnormal embryogenesis (pathogenesis)
- Manifestations of abnormal development increase in degree from the no-effect to the totally lethal level as dosage increases
- The access of adverse environmental influences to developing tissues depends on the nature of the agent
- Susceptibility to a teratogen depends on the genotype of the conceptus and on the manner in which the genotype interacts with environmental factors

Although all of these principles are important, we will limit our attention here to the first four, since they are the most relevant to epidemiologic studies of environmental reproductive hazards.

A. SPECTRUM OF ADVERSE OUTCOMES ASSOCIATED WITH EXPOSURE

The principle we want to stress is the spectrum of adverse outcomes potentially associated with exposure. Environmental studies that focus on a single outcome risk overlooking other important teratogenic (developmental) effects. Given current levels of knowledge and research methods, three outcomes should be evaluated: spontaneous abortions, congenital malformations, and low birthweight.

Low birthweight has been shown to be associated with a number of exposures, including smoking, alcohol use, and environmental chemicals such as heavy metals and toxic wastes. Spontaneous abortions are also associated with several of these exposures; for example, smoking, alcohol use, and heavy metals. Results of some occupational studies have suggested that congenital malformations and spontaneous abortions are associated with the same exposures, for example, anesthetic gases and organic solvents.

Babies with one adverse outcome may be at increased risk for another. For example, babies with congenital malformations are at increased risk for being of low birthweight. An environmental exposure may induce one or more pathogenic processes that result in more than one adverse reproductive outcome.

B. TIMING OF EXPOSURE

Wilson's second principle relates susceptibility to gestational age at time of exposure. This principle has been well established for the induction of structural malformations and the sensitive periods during development (organogenesis) are known for many structures. Considering an expanded array of outcomes, the sensitive period during which any effect can be induced must be extended throughout gestation; some would extend it throughout childhood. In designing epidemiologic studies, exposure should be determined and classified for the appropriate "time-window", that is gestational age(s), for each outcome. Spontaneous abortions and congenital malformations are likely to be related to first- and second-trimester exposure, whereas low birthweight and functional disorders are more likely to be related to second- and third-trimester exposure. It is important in epidemiologic studies to understand the biology involved when the timing of exposure is being considered.

C. TERATOGENIC MECHANISMS

The third principle is the importance of considering the potential mechanisms that might initiate abnormal embryogenesis. By considering mechanisms, investigators can develop biologically meaningful groupings of outcomes. This can also provide insight into potential teratogens; for example, relationships between carcinogenesis, mutagenesis, and teratogenesis have been discussed for some time. From our perspective, this is of particular importance for two reasons: substances that are carcinogenic or mutagenic have an increased probability of being teratogenic, suggesting that particular attention should be paid to the reproductive effects of such substances; and effects on deoxyribonucleic acid (DNA), producing somatic mutations, are mechanisms for both carcinogenesis and teratogenesis.

D. DOSE AND OUTCOME

The fourth principle concerning teratogenesis to be considered is the relationship of outcome to dose. Wilson[18] has described this principle for animal data, and Selevan and Lemasters[19] have discussed its potential relevance to the human situation. Selevan and Lemasters[19] noted the importance of multiple reproductive outcomes within specific dose ranges and suggested that a dose-response relationship could be reflected in an increasing rate of a particular outcome with increasing dose and/or a shift in the spectrum of the outcomes observed. Although one should not always expect to find a continuum of outcomes related to exposure timing and dose, these concepts provide important bases for studies.

Finally, with regard to teratogenesis and dose, we want to point out some of the problems of assessing low-dose exposures and functional disturbances. There is considerable concern about the possible neurological and behavioral effects of prenatal exposure to environmental agents. We believe that advances of knowledge in this important area depend on the development of new methods. At present, there are critical limitations in the definition and ascertainment of appropriate neurological and behavioral outcomes for epidemiologic studies.

IV. EPIDEMIOLOGIC STUDIES

Prevention of environmental health risks requires the determination of risk and the subsequent removal of, or reduction in, exposure. A key issue is the ability to determine what risks may be associated with prolonged low levels of exposure. Another issue is the ability to recognize human risks before people are exposed. In regulatory arenas there is considerable emphasis on the use of animal data to estimate human risks.[10,11] The most appropriate use of such animal data is to predict human risks before humans are exposed. When human populations are exposed then the strongest evidence for human risks comes from well-designed epidemiologic studies. The prevention of environmentally related human disease requires an expansion of knowledge regarding risks based on the combined use of appropriate human and animal data.

We now turn to a discussion of some of the adverse reproductive outcomes that should be considered when we attempt to determine if environmental contamination poses a developmental hazard. We briefly consider methods for studying four categories of adverse reproductive outcomes: spontaneous abortions, congenital malformations ("birth defects"), low birthweight, and developmental disabilities. This chapter cannot provide an encyclopedic review; the focus is on some of the basic issues regarding study approaches, rates, and methodological issues that need particular attention.

Studies of environmental reproductive hazards can be "driven" by information regarding either outcomes or exposures. That is, the impetus for study can come from a suggestion either that an adverse reproductive outcome is occurring more frequently than expected in a population or that an observed exposure is thought to increase the risk for some outcome. An example of an outcome-driven study is a study of spontaneous abortions in the area of St. Gabriel in the petrochemical corridor of Louisiana.[20] Citizens were concerned about what they considered to be excess spontaneous abortions. Examples of studies initiated on the basis of concerns about exposure are those conducted in the Love Canal area. Increasingly, epidemiologic studies are carried out in response to community concerns regarding possible exposures. Two of the critical issues that need to be addressed, particularly in the context of these studies, are the selection of reproductive outcomes to be evaluated and the determination of exposure to hazardous substances. Whatever the impetus, a study design must be developed that identifies outcomes and exposures and that analyzes the association between them.

Keep in mind that investigators (health officials) are concerned with differences between groups in the rates of occurrence for some adverse reproductive outcome. This sets up a situation with three essential features: they must be able to define the outcome, they must be able to ascertain cases of the outcome, and they must have appropriate groups for comparison — that is, populations at risk for the outcome and from which the cases have been identified. Often, problems in conducting research on environmental reproductive hazards and interpreting the results of studies have been due to problems of comparability of case definition and ascertainment between groups.

There are difficulties with which the effects of reproductive or developmental toxicants can be observed in human populations. Some of these concern the statistical frequency with which an effect can be expected to occur and the likelihood one has for observing such an effect. Much of this relates to the baseline rates of occurrence of many adverse reproductive outcomes. For example, while considerable public health concern focuses around birth defects, the rates of occurrence for most individual birth defects are quite low, less than 1/1000 births. Many studies do not have sufficient statistical power to identify any but markedly increased rates. In addition, with very few exceptions, examination of clusters of birth defects potentially thought to be associated with environmental exposures have not been rewarding in identifying a cause.

A. SPONTANEOUS ABORTIONS

In early studies of spontaneous abortions, investigators identified cohorts of women early in their pregnancies and then determined losses at various stages.[21] Results of these studies showed clearly that losses were greatest early in gestation (before 8 weeks) and that rates of loss decreased markedly as gestational age progressed. The importance of this for epidemiologic studies is clear; gestational age at pregnancy loss has an important effect on spontaneous abortion rates. When anxiety about environmental hazards increases the identification of pregnancies early in gestation, the reported number of spontaneous abortions is likely to be higher than in a nonsensitized population.

A useful approach to questions of bias in the reporting of spontaneous abortions is to compare what a woman reports with what has been recorded in previously collected records. This has been done by comparing reproductive histories reported on questionnaires with those for the same women in hospital records. In many states, a woman's previous fetal deaths are recorded on birth certificates, along with the date of the outcome. This approach has been used to examine histories of spontaneous abortion in the immediately prior pregnancy among women in different occupations.[22]

An important consideration in studies of spontaneous abortions is the ascertainment of cases. In most states, reproductive losses at less than 20 (or 28) weeks are not reportable, and, therefore, vital records cannot be used for case ascertainment. This leaves personal interviews and medical records as primary sources of case ascertainment.

Spontaneous abortions are not uncommon events; population surveys have usually reported rates in the 10 to 20% range.[23] Demographic and behavioral characteristics of the population will influence rates. This includes factors such as the maternal age distribution, the number of previous pregnancies, and smoking habits. These are in addition to one of the most important factors that influence rates, as noted above, the gestational age at which pregnancies are identified.

Spontaneous abortion is an important outcome to consider because it can result from different mechanisms and through several pathogenic processes. A spontaneous abortion can be the result of lethal embryo/fetal toxicity, chromosomal alterations, single gene effects, structural abnormalities, maternal-fetal incompatibility, or maternal abnormality. In trying to identify environmental reproductive hazards, spontaneous abortions ideally should be differentiated to four major categories: chromosomally abnormal, chromosomally normal – structurally abnormal, chromosomally normal – structurally normal, and maternal abnormality. To do this requires the collection and laboratory study of abortuses, which is not easy to accomplish.

B. CONGENITAL MALFORMATIONS

In the U.S. approximately 150,000 infants are born each year with a serious birth defect (congenital malformation). According to data from the National Center for Health Statistics, in 1992 birth defects were responsible for over 21% of all infant deaths. About 1.2 million infants, children, and adults are hospitalized each year for treatment of birth defects. Of considerable public health significance is the fact that the causes of birth defects in humans are largely unknown. Along with eliminating or reducing exposures to known teratogens, attention must focus on identifying environmental factors that may contribute to the unknown portion of the spectrum.

One of the greatest public concerns about potential health effects of environmental exposures relates to possible associations with birth defects or congenital malformations. Even though this concern may be well founded, given animal studies that suggest adverse effects of some environmental chemicals and a recognition of the susceptibility of the human embryo to extrinsic substances, convincing data linking birth defects with environmental exposures are lacking. As noted above, the major exception to this is the association of mental retardation and cerebral palsy with methylmercury contamination in Minimata Bay, Japan. Here, however, the association was with developmental defects — that is functional disorders, rather than with congenital malformations — that is, structural defects.

One advantage to the study of congenital malformations is that baseline rates have been established for many defects. This allows the investigator to compare an observed rate with what is expected. The presence of birth defects surveillance programs in many states has led to the development of rates for these areas. Data for all parts of the country are available from the Birth Defects Monitoring Program, operated by the Centers for Disease Control and Prevention.

There is growing interest in the development of birth defects surveillance programs as a means of identifying potential environmental reproductive hazards. Projects have been carried out in several states linking data from surveillance programs with information on environmental contamination from environment databases. Examples include a study of birth defects and proximity to toxic waste sites in New York[24] and studies of water contaminants and birth defects in New Jersey and California.[25,26]

An important similarity between birth defects and spontaneous abortions is the impact of methods of case ascertainment on rates. Relatively few investigators have examined the effect on rates of using different sources of ascertainment. Historically important studies showed the marked under-reporting of many malformations on vital records compared with those malformations recorded in hospital records. Even for such readily observable congenital malformations as the neural tube defects anencephaly and spina bifida, rates differ by methods of case ascertainment. The importance of this relates to questions regarding expected rates based on surveillance programs.

Current surveillance programs employ a variety of case ascertainment methods and data sources. Systems employing active surveillance methods and multiple sources of case ascertainment are more likely to approximate true prevalence at birth ("incidence") than are passive systems based primarily on vital records.

The occurrence of a cluster of cases of anencephaly in Brownsville, Texas, has led to recommendations that birth defects surveillance be expanded. Much of the initial impetus for birth defects surveillance programs grew out of concerns regarding the early identification of newly introduced teratogens.[27]

Clusters of birth defects present particular problems. Most studies of birth defect clusters have been unrewarding; no cause has been found. We should, however, be aware of the fact that the major human teratogens, rubella virus[28] and thalidomide,[29] were recognized by alert clinicians who observed excesses of particular defects among their patients. Birth defect clusters present a conundrum; if we do not investigate them, we may miss another important teratogen, but, at the same time, we are not likely to identify any etiology.

In summary, congenital malformations are an important outcome for examining potential reproductive effects of environmental exposures. The low rates of individual congenital malformations often lead to low statistical power of malformation studies. This creates a need to aggregate malformations in a biologically meaningful way for analysis. For studies of malformations, rates based on multiple sources of ascertainment can be compared. The increase in state-based birth defects surveillance programs provides opportunities for additional information on expected rates for geographical regions of the country. This information will be used increasingly to identify, and hopefully to resolve, public health problems.

C. BIRTHWEIGHT

The next reproductive outcome considered is birthweight, particularly birthweight distributions. Birthweight may be a sensitive indicator of effects of environmental exposures on reproductive outcome. Low birthweight or a depressed birthweight distribution may be an indicator of intrauterine growth retardation. Birthweight has at least five advantages for study: (1) it is routinely collected through the vital records system, (2) it can be studied as either a continuous or categorical variable, (3) the exposure time window for birthweight is not as restricted as for congenital malformations, (4) studies have clearly shown effects of exogenous agents on birthweight, and (5) dose-response relationships have been demonstrated.

In studying potential associations between environmental exposure and birthweight, I believe that the most important focus is on growth retardation as opposed to prematurity. Therefore, information on gestational age is needed in order to determine gestational age-specific birthweight distributions.

A major question is whether existing vital records databases can be used to study birthweight distributions, given potential confounders. Ways of describing populations on the basis of socioeconomic variables have been explored. The collection of information on smoking status, as part of the birth certificate, is an important step in making these data more useful for epidemiologic studies.

D. DEVELOPMENTAL DISABILITIES

The last outcome I want to consider briefly is the broad category of developmental disabilities. This category of functional abnormalities includes an array of possible endpoints such as cerebral palsy, mental retardation, and severe seizure disorders at one end of the spectrum, with more subtle intellectual impairment, behavioral abnormalities, and learning disabilities at the other. For the more subtle outcomes, our knowledge of baseline rates tends to be limited and there are problems with case definition and assessment at the population level. At the same time, the strongest evidence for an effect of environmental contamination on human development comes from methylmercury exposure leading to neurodevelopmental effects. In addition, one of the most important human teratogens, alcohol, has mental retardation as its most striking effect and much of the concern about lead in the environment focuses on its possible cognitive effects. Additional attention needs to be paid to human behavioral/functional teratology and to the development of epidemiologic methods for studying these effects.

E. EPIDEMIOLOGIC STUDIES SUMMARY

In conclusion, I have reviewed four major categories of adverse reproductive outcomes potentially associated with environmental teratogenesis. These outcomes range from spontaneous abortion and congenital malformations, to retarded intrauterine growth expressed as a reduction in birthweight, to functional disorders manifested as developmental disabilities. Although the relationships between these outcomes are not clear, it is plausible that a given environmental exposure could have varying effects, depending on exposure timing and dose. It is essential that epidemiologists pay more attention to principles of teratogenesis in studying potential environmental reproductive hazards and in designing appropriate studies.

V. EFFECTS OF SPECIFIC AGENTS ON DEVELOPMENT

We shall now review the effects of prenatal exposure to certain environmental agents. Agents were selected on the basis of the availability of human data and public concerns that the agent may cause adverse reproductive effects. Several texts and reviews provide more comprehensive information about the potential teratogenicity of numerous environmental agents and chemicals.[2,3,30,31]

Known human teratogens number about 30. Most of these are drugs and, to date, only two teratogens have been shown conclusively to cause human maldevelopment (mental retardation and cerebral palsy) as a result of environmental contamination: high-dose ionizing radiation and methylmercury. The example of high-dose radiation is unique to *in utero* exposure following the atomic bombings of Hiroshima and Nagasaki. Agents suspected of being teratogens on the basis of animal studies are more numerous, and when these agents are found to be polluting the environment, there is justifiable concern about potential prenatal exposure and adverse reproductive effects. As was discussed in considering epidemiologic studies, there are difficulties in demonstrating an association between an environmental hazard and any adverse reproductive outcome.

A. METHYLMERCURY

Exposure to methylmercury is not common. Methylmercury was widely used as a fungicide in the U.S. until the 1960s when production was halted because of the compound's toxicity and ability to bioaccumulate. In the U.S. today exposure occurs through the consumption of contaminated fish.

The adverse effects of methylmercury on the fetus were discovered in 1959. There was an epidemic of poisoning in Minamata, Japan, that resulted in fetal neurologic damage with psychomotor retardation, seizures, cerebral palsy, blindness, and deafness.[32] In 1972, similar fetal effects were observed in Iraq after an epidemic of methylmercury poisoning associated with the use of mercury-treated seed grain for flour.[33]

The developing nervous system of the conceptus and infant is more sensitive to the toxic effects of organic mercury than that of the adult or older child. The devastating neurological effects become evident only as abnormal neurologic development proceeds. Methylmercury does not cause obvious structural malformations in humans.

B. HEAVY METALS

Since the 19th century exposure to high levels of lead has been known to cause embryotoxicity, growth retardation, increased perinatal mortality, and developmental disability including mental retardation.[34]

Residents of communities near metal smelters may be exposed to high concentrations of lead, as well as to other heavy metals, in air through smelter emissions. A series of ecologic studies from Sweden showed an association between residential proximity to a copper smelter and increased rates of spontaneous abortion and low birthweight.[35,36] Air or blood metal concentrations were not measured and risk factors such as socioeconomic status, maternal age, and parity also were not considered in the analysis, so the association may be questioned. In addition, there is evidence to suggest that more of the women who lived near the smelters were employed there and were at risk for occupational exposures.

Industrial waste, copper smelter emissions, and some pesticides are sources of environmental pollution with arsenic. Decreased birthweight and increased spontaneous abortion rates associated with residential proximity to the copper smelter in Sweden may have resulted from arsenic exposure, since arsenic, lead, and other heavy metals are present in copper smelter emissions.[3]

C. AGRICULTURAL CHEMICALS AND PESTICIDES

Concerns have been expressed about the reproductive effects of a variety of pesticides and other agricultural chemicals.[37] As with many environmental health issues, the concern is based not so much on convincing data as on the biological plausibility of such effects.

Some ecological studies have suggested associations between agricultural chemical usage and specific congenital malformations, such as neural tube defects,[38] limb-reduction defects,[39] and cleft lip and palate.[40] Other studies have suggested associations between pesticides and spontaneous abortions. While some of these ecological studies have been supported by epidemiologic studies that include the assessment of individual exposures, for the most part questions of reproductive and developmental risks remain unresolved.

Major substances of concern have been the phenoxy herbicide 2,4,5-T and its contaminant dioxin.[3] These compounds are considered together because 2,3,7,8-tetrachlorodibenzodioxin (TCCD or dioxin) and other chlorinated dibenzodioxins were produced as contaminants during production of the herbicide, 2,4,5-trichlorophenoxyacetic acid (2,4,5-T). This herbicide is no longer marketed in the U.S., largely because of concerns about possible teratogenic and embryotoxic effects.

Agent Orange, a defoliant used in Vietnam, was a mixture of herbicides, including 2,4,5-T. During the later years of the Vietnam war, public opinion against the ecological effects of Agent Orange was fueled by reports of birth defects in South Vietnamese babies born to mothers who lived in areas where Agent Orange had been sprayed. The ensuing debate about the potential human teratogenicity of Agent Orange involved numerous federal agencies, and the use of 2,4,5-T containing any chlorodioxin contaminants was cancelled in 1979.[41]

D. NITRATES IN WATER

Concerns have been raised about developmental effects of nitrates in drinking water. Nitrate contamination of water supplies may result from agricultural (fertilizer) run-off, sewage, or industrial waste.

Epidemiologic studies do not provide conclusive evidence that pregnant women who consume low levels of nitrates from drinking water are at increased risk for having adverse reproductive outcomes. In South Australia an excess of birth defects led to a case-control study examining the relationship between maternal drinking water source (groundwater vs. rainwater) and risk of congenital malformations.[42] The risk of having a malformed infant was increased among women who drank groundwater. Risks for central nervous system malformations and oral clefts were particularly increased. A dose-response relationship was found using estimated nitrate concentrations. Strengths of the study include the completeness of case ascertainment and the monitoring of water nitrates during the study period. Limitations include the assumption that water concentrations were constant during monitoring intervals and that the subjects used the same drinking water source throughout pregnancy. Another assumption is that nitrates rather than some unmeasured drinking water contaminant was responsible. There could be unrecognized confounding by a third variable. For example, seasonal variation in malformation risks suggests that dietary, nutritional, or other environmental factors may have contributed to the increased malformation rates.

A Canadian case-control study found an increased risk for delivering an infant with a central nervous system malformation associated with exposure to nitrates through water from private wells.[43] The opposite was found with drinking water obtained from other sources. To assess exposure, the investigators analyzed nitrates in water samples collected at addresses where study subjects lived at the time of delivery. This raises the important issue of determining exposures at relevant stages in pregnancy; recent studies have shown a high degree of residential mobility during pregnancy. The study also is limited by the lack of information about other possible water contaminants. The opposite risks associated with drinking water source, independent of nitrate concentration, also suggest that other factors contributed to the observed effects.

E. POYLCHLORINATED BIPHENYLS

Polychlorinated biphenyls (PCBs) had numerous industrial uses because they are heat stable and transfer heat effectively. In addition, PCBs are extremely resistant to both metabolic and biologic degradation. Because of past dumping or disposal in unregulated landfills and failure to recycle, PCBs have become ubiquitous in the environment.

Low-level maternal exposure to PCBs can occur throughout life, by consuming fish from PCB-contaminated waters as well as other dietary sources. While placental transfer of PCBs occurs, the largest dose is delivered to the nursing infant via breast milk.[44] The widespread potential for maternal exposure and PCB transfer to nursing infants has resulted in justifiable concern about developmental effects.

In epidemiologic studies, maternal exposure to PCBs via fish consumption is associated with a slight decrease in infant birthweight and head circumference, compared to babies of unexposed controls.[45] Neonatal assessment of the same infants provides some evidence that at low concentrations, PCBs may be behavioral teratogens. As discussed earlier, behavioral abnormalities may reflect functional impairment caused by prenatal exposure to a chemical agent, such as PCBs. In infants studied at 4 years of age, short-term memory impairment was directly related to the umbilical cord serum PCB concentration.[46]

F. POLYBROMINATED BIPHENYLS

Polybrominated biphenyl (PBB) compounds are structurally similar to PCBs and they share characteristics of high fat solubility and resistance to degradation. PBBs are not as widespread environmental contaminants as PCBs because their uses have been more limited.

Given that PCBs are behavioral teratogens and that PBBs are structurally and chemically similar to PCBs, it is logical to consider the possibility of similar neurobehavioral effects from prenatal and infant PBB exposure. An early follow-up study reported that children with higher PBB body burdens scored lower on standardized tests of perceptual motor, attentional, and verbal abilities. Subsequent testing of these children showed that their overall developmental scores were within the normal range. For certain perceptual and perceptual-motor tasks, the scores tended to be inversely related to PBB body burden.[47]

G. ORGANIC SOLVENTS

On numerous occasions, organic solvents in drinking water have resulted from storage tank leaks or contamination from hazardous waste sites. Such contamination is a common issue at designated "Superfund" sites since over half report the presence of organic solvents. Affected communities may identify a temporal or geographic clustering of adverse reproductive effects, believing that the contamination has caused the epidemic. To date, there is no published evidence to support such a cause-effect relationship, but at least one study demonstrated an increased risk of spontaneous abortions among women during the time that their drinking water supply was contaminated with trichloroethane and other organic solvents. A hospital-based study also was undertaken in response to community concerns in California that increased numbers of children with congenital heart defects were born to women living in the contaminated area.[48] The increased prevalence of heart defects was confirmed but found to be temporally unrelated to the drinking water contamination, making a causal relationship unlikely.

A more recent study in Arizona has suggested an association between trichlorethylene (TCE) in water and heart malformations.[49] TCE exposure is particularly wide spread and an investigation of water contamination and malformations in Woburn, Massachusetts, did not show an association with cardiac defects, although risks of some malformations were reported to be elevated.[50] Exhaustive studies of malformations in Woburn, a community that has achieved considerable notoriety for water contamination and adverse health effects, including adverse reproductive outcomes, were compiled recently and do not show increased congenital malformation risks.[51]

A series of reports from the New Jersey Department of Health has suggested associations between several organic solvents and specific congenital malformations.[25] Based on cases of malformations from the New Jersey Birth Defects Registry, and databases with information on water contamination, a number of statistically significant associations were demonstrated. Additional studies are being carried out using more refined methods of exposure assessment.

There is no conclusive evidence that low-level exposures to organic solvents increase the risk of congenital malformations or other adverse reproductive outcomes. However, because there are suggestive data and because organic solvents are such widespread environmental contaminants, additional epidemiologic studies of a variety of reproductive outcomes in exposed populations are indicated.

VI. CONCLUSIONS

The topic of potential hazards to reproduction associated with environmental contamination is of public health significance. Epidemiologic studies can play an important role in attempting to identify existing hazards and can contribute to the process of risk assessment. In studying possible hazards, it is essential to keep the principles of teratogenesis in mind and to design studies that consider an array of possible outcomes. Substances of particular concern that require further investigation include agricultural chemicals, such as pesticides, and organic solvents. All such studies require an ability to adequately assess exposure and the potential role of biological markers of exposure needs to be exploited fully.

REFERENCES

1. Postel, S., *Defusing the Toxics Threat: Controlling Pesticides and Industrial Waste*, Worldwatch Institute, Washington, D.C., 1987.
2. Clement Associates, *Chemical Hazards to Human Reproduction*, U.S. Government Printing Office, Washington, D.C., 1981.
3. Mortensen, M. E., Sever, L. E., and Oakley, G. P., Jr., Teratology and the epidemiology of birth defects, in *Obstetrics: Normal and Problem Pregnancies*, 2nd ed., Gabbe, S. G., Niebyl, J. R., and Simpson, J. L., Eds., Churchill Livingstone, New York, 1991, 233.
4. Sever, L. E. and Hessol, N. A., Overall design considerations in male and female occupational reproductive studies, in *Reproduction: The New Frontier in Occupational and Environmental Health Research*, Lockey, J. E., Lemasters, G., and Keye, W. R., Jr., Eds., Alan R. Liss, New York, 1984, 15.
5. Sever, L. E. and Hessol, N. A., Toxic effects of occupational and environmental chemicals on the testes, in *Endocrine Toxicology*, Thomas, J. A., McLachlan, J. A., and Korach, K. S., Eds., Raven Press, New York, 1985, 211.

6. Wyrobek, A. J., Watchmaker, G., and Gordon, L., An evaluation of sperm tests as indications of germ-cell damage in men exposed to chemical or physical agents, in *Reproduction: The New Frontier in Occupational and Environmental Health Research*, Lockey, J. E., Lemasters, G., and Keye, W. R., Jr., Eds., Alan R. Liss, New York, 1984, 385.
7. Levine, R. J., Methods for detecting occupational causes of male infertility, *Scand. J. Work Environ. Health*, 9, 371, 1983.
8. National Research Council, *Biologic Markers in Reproductive Toxicology*, National Academy Press, Washington, D.C., 1989.
9. Baird, D. D., Wilcox, A. J., and Weinberg, C. R., Use of time to pregnancy to study environmental exposures, *Am. J. Epidemiol.*, 124, 470, 1986.
10. U.S. Environmental Protection Agency (EPA), Proposed guidelines for assessing female reproductive risk and request for comments, *Fed. Reg.*, 53, 24834, 1988.
11. U.S. Environmental Protection Agency (EPA), Proposed guidelines for assessing male reproductive risk and request for comments, *Fed. Reg.*, 53, 24850, 1988.
12. Fabia, J. and Thuy, T. D., Occupation of father at time of birth of children dying from malignant disease, *Br. J. Prev. Soc. Med.*, 28, 98, 1974.
13. Gardner, M. J., Snee, M. P., Hall, A. J. et al., Results of a case-control study of leukaemia and lymphoma among young people near Sellafield Nuclear Plant in West Cumbria, *Br. Med. J.*, 300, 423, 1990.
14. Spitz, M. and Johnson, C., Neuroblastoma and paternal occupation: a case control analysis, *Am. J. Epidemiol.*, 121, 924, 1985.
15. Savitz, D., Childhood cancer, *Occup. Med.*, 1, 415, 1986.
16. Sever, L. E., Parental radiation exposure and children's health: are there effects on the second generation?, *Occup. Med.*, 6, 613, 1992.
17. Hatch, M., The epidemiology of electric and magnetic field exposures in the power frequency range and reproductive outcomes, *Pediatr. Perinat. Epidemiol.*, 6, 198, 1992.
18. Wilson, J. G., *Environment and Birth Defects*, Academic Press, New York, 1973.
19. Selevan, S. G. and Lemasters, G. K., The dose-response fallacy in human reproductive studies of toxic exposures, *J. Occup. Med.*, 29, 451, 1987.
20. White, L. E., Mather, F. J., and Clarkson, J. R., Final Report St. Gabriel Miscarriage Investigation, East Bank of Iberville Parish, Louisiana, Louisiana Department of Health and Hospitals, Baton Rouge, LA, 1989.
21. Taylor, W. F., On the methodology of measuring the probability of fetal death in a prospective study, *Hum. Biol.*, 36, 86, 1964.
22. Vaughan, T. L., Daling, J. R., and Starzyk, P. M., Fetal death and maternal occupation, *J. Occup. Med.*, 26, 676, 1984.
23. Hertz-Picciotto, I. and Samuels, S. J., Incidence of early loss of pregnancy, *N. Engl. J. Med.*, 319, 1483, 1988.
24. Geschwind, S. A., Stolwijk, J. A. J., Bracken, M. et al., Risk of congenital malformations associated with proximity to hazardous waste sites, *Am. J. Epidemiol.*, 135, 1197, 1992.
25. Bove, F. J., Fulcomer, M. C., Klotz, J. B. et al., Report on Phase IV-B: Public Drinking Water Contamination and Birth Weight and Selected Birth Defects — A Case-Control Study, New Jersey Department of Health, 1992.
26. Shaw, G. M., Schulman, J., Frisch, J. D., Cummins, S. K., and Harris, J. A., Congenital malformations and birthweight in areas with potential environmental contamination, *Arch. Environ. Health*, 47, 147, 1992.

27. Edmonds, L. D., Layde, P. M., James, L. M. et al. Congenital malformations surveillance: two American systems, *Int. J. Epidemiol.*, 10, 247, 1981.
28. Gregg, N. M., Congenital cataract following german measles in the mother, *Trans. Ophthalmol. Soc. Aust.*, 3, 35, 1941.
29. Lenz, W., Discussion Contribution by Dr. W. Lenz, Hamburg, on the Lecture by R.A. Pfeiffer and K. Kosenow: On the Exogenous Origin of Malformations of the Extremities, Tagung der Rheinisch-Westfalischen Kinderarztevereinigung, Dusseldorf, 1961.
30. Schardein, J. L., *Chemically Induced Birth Defects.*, Marcel Dekker, New York, 1985.
31. Shepard, T. H., *Catalog of Teratogenic Agents*, 6th ed., Johns Hopkins University Press, Baltimore, 1989.
32. Harada, M., Minamata disease: a medical report, in *Minamata*, Smith, W. E. and Smith, A. M., Eds., Holt, Rinehart, New York, 1975, 180.
33. Bakir, F., Damluji, S. F., Amin-Zaki, M. et al., Methylmercury poisoning in Iraq, *Science*, 181, 230, 1973.
34. Rom, W. N., Effects of lead on the female and reproduction: a review, *Mount Sinai J. Med.*, 43, 542, 1976.
35. Nordstrom, S., Beckman, L., and Nordenson, I., Occupational and environmental risks in and around a smelter in Northern Sweden. II. Chromosomal aberrations in workers exposed to arsenic, *Hereditas*, 88, 47, 1978.
36. Nordstrom, S., Beckman, L., and Nordenson, I., Occupational and environmental risks in and around a smelter in Northern Sweden. III. Frequencies of spontaneous abortion, *Hereditas*, 88, 51, 1978.
37. Moses, M., Pesticide-related health problems and farm workers, *AAOHN J.*, 37, 115, 1989.
38. White, E. M. M., Cohen, F. G., Silverman, G., and McCurdy, R., Chemicals, birth defects and stillbirths in New Brunswick: associations with agricultural activity, *Can. Med. Assoc. J.*, 138, 117, 1988.
39. Schwartz, D. A. and LoGerfo, J. P., Congenital limb reduction defects in the agricultural setting, *Am. J. Public Health*, 78, 654, 1988.
40. Gordon, J. R. and Shy, C. M., Agricultural chemical use and congenital cleft lip and/or palate, *Arch. Environ. Health*, 26, 213, 1981.
41. U.S. Environmental Protection Agency (EPA), 2,4,5-T and Silvex, *Fed. Reg.*, 44, 15874, 1979.
42. Dorsch, M. M., Scragg, R. K. R., McMichael, A. J. et al., Congenital malformations and maternal drinking water supply in rural South Australia: a case-control study, *Am. J. Epidemiol.*, 119, 473, 1984.
43. Arbuckle, T. E., Sherman, G. J., Corey, P. H. et al., Water nitrates and CNS birth defects: a population-based case-control study, *Arch. Environ. Health*, 43, 162, 1988.
44. Schwartz, P. M., Jacobson, S. W., Fein, G. et al., Lake Michigan fish consumption as a source of polychlorinated biphenyls in human cord serum, maternal serum, and milk, *Am. J. Public Health*, 73, 293, 1983.
45. Fein, G. G., Jacobson, J. L., Jacobsen, S. W. et al., Prenatal exposure to polychlorinated biphenyls: effects on birth size and gestational age, *J. Pediatr.*, 105, 315, 1984.
46. Jacobsen, J. L., Jacobson, S. W., and Humphrey, H. E., Effect of *in utero* exposure to polychlorinated biphenyls and related contaminants on cognitive functioning in young children, *J. Pediatr.*, 116, 38, 1990.
47. Schwartz, E. M. and Rae, W. A., Effect of polybrominated biphenyls (PBB) on developmental abilities in young children, *Am. J. Public Health*, 73, 277, 1983.

48. Swan, S. H., Shaw, G., Harris, J. A. et al., Congenital cardiac anomalies in relation to water contamination, Santa Clara County, California, 1981-1983, *Am. J. Epidemiol.*, 129, 885, 1989.
49. Goldberg, S. J., Lebowitz, M. D., and Graver, E. J., An association of human congenital cardiac malformations and drinking water with contaminants, *J. Am. Coll. Cardiol.*, 16, 155, 1990.
50. Lagakos, S. W., Wessen, B. J., and Zelen, M., An analysis of contaminated well water and health effects in Woburn, Massachusetts, *J. Am. Stat. Assoc.*, 81, 583, 1986.
51. Massachusetts Department of Public Health, Woburn Environment and Birth Study, 1994.

CHAPTER 7

Risk Factors for Cancer in the Occupational Environment and Relevant Epidemiologic Study Methods

Jack Siemiatycki

I. INTRODUCTION

Cancer is a general term for a group of related diseases characterized by the uncontrolled growth of certain tissues. This chapter concerns environmental causes of cancer, and in particular, the chemical and physical agents in the occupational environment that cause cancer. For reasons to be elaborated below, the occupational milieu has been the most fertile ground for studying the effects of environmental chemical and physical agents. This chapter is intended to give the nonspecialist reader a brief historical and conceptual overview of the problem of environmental cancer, showing in part how limited our knowledge is and how important epidemiology is in filling the knowledge gap. Then, we will consider how various epidemiologic designs can be used for the purpose of investigating occupational carcinogens.

II. HISTORY

Although cancer has been recognized since antiquity, until recently its etiology was a complete mystery. Over the past century, there has been growing recognition that cancer could have an environmental origin. In the 18th century there were some scattered, and largely ignored, reports implicating tobacco in the form of snuff in nasal cancer[1] and in the form of cigarettes in lip cancer.[2] In 1775, Sir Percival Pott, a British surgeon, published a brief essay in which he ascribed some cases of cancer of the scrotum among chimney sweeps to the pitifully dirty working conditions. More specifically, he ascribed this condition, which was known in the trade as "soot

0-87371-573-X/95/$0.00+$.50

wart", to the "lodgment of soot in the rugae of scrotum".[3] Pott's hypothesis was based on astute clinical observation rather than on rigorous scientific methods. In the ensuing century, the syndrome became widely known, but it remained the only recognized environmentally caused cancer until the latter part of the 19th century. In 1875, Volkmann described a syndrome identical to "chimney sweeps' cancer" of the scrotum among a group of coal tar and paraffin workers.[4] Apparent clusters of scrotal cancer were thereafter reported among shale oil workers[5] and mule spinners in the cotton textile industry.[6,7] By 1907 the belief in the carcinogenicity of "pitch, tar and tarry substances" was widespread enough that skin cancers among exposed workers were officially recognized as compensable in the United Kingdom. Other types of cancer were also implicated as occupationally induced. In 1879, Harting and Hesse reported that pulmonary cancer was common among metal miners.[8] In 1895, Rehn reported that he was struck by the number of bladder cancer cases diagnosed among workers from a local plant which produced dyestuffs from coal tar.[9]

The scientific investigation of cancer etiology began in earnest at the beginning of the 20th century with experimental animal research. Among the important early experimental findings, it was shown that cancer was not a transmissible disease, such as smallpox, cholera, and the like, and it was shown that in genetically mixed populations, genetic linkage was not an important determinant of cancer risk. A major breakthrough came with the experiments of Yamagiwa and Ichikawa in which they showed that coal tar applied to rabbit ears produced skin tumors.[10] Several important discoveries were made in the next 20 years. In a series of experiments, an English group led by Kennaway isolated dibenz(*a*,*h*)anthracene and benzo(*a*)pyrene, both polycyclic aromatic hydrocarbons (PAHs) and active ingredients in coal tar.[11-13] Several other PAHs were subsequently shown to be carcinogenic to laboratory animals, but so were substances of many other chemical families. For instance, 2-naphthylamine was shown to cause bladder tumors in dogs and this was thought to explain the bladder cancers seen earlier among dyestuffs workers.

Between 1930 and 1960, there were additional reports of high-risk occupational groups, but there was no clear unifying thread. Respiratory cancer risks were reported in such diverse occupational settings as nickel refineries,[14] coal carbonization processes,[15] chromate manufacture,[16] manufacture of sheep-dip containing inorganic arsenicals,[17] gas workers,[18] and asbestos workers.[19,20] Workers producing dyestuffs in the chemical industry were found to have excess bladder cancer.[21]

By the 1940s there was growing suspicion that tobacco products could cause cancers of the buccal cavity and of the lung. The era of modern epidemiologic studies of environmental cancer began around 1950 with the publication of five studies of smoking and lung cancer.[22-26] These studies were unanimous in concluding that smoking was a powerful risk factor for lung cancer. Because of the widespread nature of the smoking habit and because of the high risk of cancer that this habit confers, it has been shown that a very large fraction of cancer could be prevented if smoking were eliminated.[27] No more effective preventive measure is known and consequently, in terms of potential public health impact, the findings of these epidemiologic studies on smoking and lung cancer remain the most important results not only in the field of cancer epidemiology but probably in all of cancer research. These studies were also important in drawing attention and legitimacy to the case-control approach to studying the effects of risk factors in epidemiology.

Subsequently, and especially with the flowering of "environmentalism" in the 1960s as a component of social consciousness, there has been a sharp increase in the amount of research aimed at investigating links between the environment and cancer. For several reasons, there has been a particular emphasis on the occupational environment. Most of the historic observations of environmental cancer risks were discovered in occupationally exposed populations. As difficult as it is to characterize and study groups of workers, it is much harder to study groups of people who share other characteristics, such as diet or general environmental pollution. Not only are working populations easier to delineate but, often, company personnel records and more recently industrial hygiene records permit some, albeit crude, form of quantification of individual workers' exposure to workplace substances. Also, the pressure of organized labor has been an important force in attracting attention to the workplace. Finally, the workplace is a setting where people have been exposed to high levels of many substances that could potentially be harmful. Since many occupational exposures can also occur in the general environment, the cancer risks borne by workers have implications well beyond the workplace.

Chemical and physical agents in the workplace represent a major class of environmental risk factors for cancer, but they are not the only ones. In the past 40 years, epidemiologic research has been instrumental in demonstrating that various human environments and activities may influence the risk of cancer. These include personal habits, such as smoking and alcohol consumption; aspects of dietary intake, such as amount of fiber and various vitamins; reproductive/sexual behavior, such as age at first birth and sexual promiscuity; medication use, such as estrogen replacement therapy and immunosuppressive anticancer drugs; microbiological agents, such as certain viruses and parasites; ionizing radiation from nuclear bombs; and aspects of the "natural" terrestrial environment, such as sunlight.[28]

For a variety of reasons, it is reasonable to surmise that the environment, taken in its broadest sense to include all of the elements listed above and any other physical/ social/ psychological/cultural phenomena that surround and interact with us on the journey from cradle to grave, has a major impact on the risk of cancer. Some empirical evidence to support this contention includes the following: (1) there is tremendous variation in cancer risks among countries and even among subnational units, (2) within communities, the incidence of cancer can change dramatically in the space of a few decades, (3) the pattern of cancer incidence changes in migrants from that prevalent in their native land to something approaching that of the host country, (4) there are now many documented examples, such as those listed above, of specific risk factors for cancer which in aggregate can be shown to contribute to a large number of cases. There have been claims that upwards of 80 or 90% of cancers are attributable to environmental, as opposed to genetic, factors. But this formulation is based on a false dichotomy. More likely, all cases are attributable to both genetically determined susceptibility and environmental risk factors of some sort.

III. CHARACTERISTICS OF CARCINOGENS

A carcinogen may be defined mechanistically as a substance that causes cancer under certain, perhaps poorly understood, circumstances, or it may be defined probabilistically as a substance that increases the risk of developing a tumor among those subjects who are exposed. A carcinogen may contribute to the occurrence of a tumor that would not otherwise have occurred or it may hasten the onset of a tumor. Exposure to a carcinogen does not usually make cancer inevitable. For instance, cigarette smoking increases the lifetime risk of lung cancer among North American males from about 1% among nonsmokers to about 10% among smokers, and cigarette smoke is thus considered to be a powerful lung carcinogen. Nevertheless, 90% of smokers will not get lung cancer.

The biologic mechanisms of carcinogenesis are complex and are not yet fully understood. Detailed reviews can be found in many sources.[29-31] For our purpose it is useful to mention a few principles that influence the methodology of epidemiologic studies in this area. Cancers are multifactorial diseases, both in a population sense and in an individual sense. At the population level, a given type of cancer can have different causes among different people. At the individual level, each case results from the unfortunate conjuncture of a complex of circumstances, e.g., genetic predisposition, diet, environmental pollutant, occupational exposure, medical intervention, viral infection, lifestyle habits, etc., which combine over a lifetime to initiate and promote the tumor growth. Each of the contributory factors may be considered to be a cause even though it is not sufficient to cause the disease by itself. Carcinogenesis appears to be a multistage process; some carcinogens act early in the process, while others act late. There may be a long induction period between exposure and appearance of the tumor; thus, the search for carcinogenic exposures in a subject's history must account for a long part of a lifetime. Interactions between specific chemicals sometimes act to enhance the carcinogenic process and sometimes to inhibit it. With the possible exception of some viral-induced cancers, it is not possible to infer the cause of a tumor by examining the tissue. That is, a chemically or physically-induced tumor bears no definitive trace of its cause. There may be traces of substances to which the person was exposed and which may have caused the tumor, as in the examples of asbestos fibers in lung tissue or DNA adducts of chemical carcinogens; at this time, however, we have no way of telling whether such indicators were in fact causes of the tumor, innocent by-standers of the process, or sequellae of the cancerous process. Most known carcinogens affect only one or a small number of bodily organs; thus, the investigation of carcinogens must be carried out on a cancer site by cancer site basis.

There is tremendous variability in the carcinogenic potency of different carcinogens. The degree of risk due to exposure to a given carcinogen increases with increased dose, but the shape of the dose-response curve may differ from one carcinogen to another. Depending on the biological fate of the substance in the body and other factors, the risk of cancer may be a function of the total exposure to the substance accumulated over a lifetime or it may be a function of exposure accumulated in a window of time. Some carcinogens can act in a short exposure period or even a single high dose; others require prolonged exposure. Some act directly on the cells; others require metabolic transformation to be carcinogenic. Some carcinogens appear to exert their effect by virtue of their mutagenic properties (i.e., their ability to irreversibly alter the genetic material of cells), whereas others appear to have different mechanisms, such as depressing immunologic function or promoting tumor growth. Cancer risk is determined by the exposure circumstance and by the host susceptibility. The exposure circumstance includes such parameters as the dose, the duration, and the concomitant conditions or exposures. Host susceptibility concerns such parameters as metabolic activation rates, genetic repair mechanisms, hormonal and immunologic profiles.

IV. LISTING OCCUPATIONAL CARCINOGENS

Although it seems like a simple enough task, it is very difficult to draw up an unambiguous list of occupational carcinogens. Different criteria would lead to different ones being listed. In some instances, we know that an occupational or industrial group is at excess risk of cancer and we have a good idea of the causative agent (e.g., lung cancer among asbestos miners and asbestos fibers.[32] In some instances, we know that a group experienced excess risk but the causative agent is unknown or at least unproven (e.g., nasal sinus cancer among furniture workers).[33] The strength of the evidence for an association can vary. For some associations the evidence of excess risk seems incontrovertible (e.g., liver cancer and vinyl chloride monomer).[34] For some associations the evidence is only suggestive (e.g., lung cancer and diesel engine exhaust).[35] Among the many substances in the industrial environment for which there are no human data concerning carcinogenicity, there are hundreds that have been shown to be carcinogenic in some animal species and thousands that have been shown to have some effect in assays of mutagenicity.

All of these considerations at best complicate and at worst tend to undermine the legitimacy of any attempt to devise a list of recognized carcinogens. Nevertheless it is necessary to put some order in our knowledge and the drafting of such lists is useful. For this task we drew largely on the Monograph Programme of the International Agency for Research on Cancer, the objective of which is to publish critical reviews of epidemiologic and experimental data on carcinogenicity for chemicals, groups of chemicals, industrial processes, and other complex mixtures to which humans are known to be exposed.[36]

Table 1 shows those chemical and physical agents in the workplace for which there is at least some strong suggestion of an association with human cancer. This table includes substances for which there are studies in humans supporting the association, sometimes with additional support from animal studies. We have indicated the type(s) of cancer affected, with an indication of the strength of evidence for each type listed. Table 1 excludes substances that have been shown to be animal carcinogens but for which there is no corroboration in humans, and it also excludes substances to which there might be occupational exposure, but whose principal mode of exposure is not related to occupation, such as tobacco, hormones, and pharmaceu-

Table 1 Substances that are Recognized or Probable Occupational Carcinogens, with an Indication of the Types of Cancer Affected and the Strength of the Evidence

Substance	Site	Strength of the Evidence
Aromatic amines		
2-Naphthylamine	Bladder	Strong
4-Aminobiphenyl	Bladder	Strong
Benzidine	Bladder	Strong
Mixtures containing PAHS		
Coal tars and pitches	Skin	Strong
	Lung	Moderate
	Bladder	Moderate
Creosotes	Skin	Moderate
Diesel engine exhaust	Lung	Moderate
	Bladder	Moderate
Mineral oils, untreated and mildly treated	Skin	Strong
	Nasal sinuses	Moderate
	Respiratory tract	Moderate
	Bladder	Moderate
Shale oils	Skin	Strong
Soots	Skin	Strong
	Lung	Strong
	Esophagus	Moderate
Other organics		
Acrylonitrile	Lung	Moderate
	Prostate	Moderate
Bis-chloromethyl ether	Lung (oat cell)	Strong
Benzene	Leukemia	Strong
Ethylene oxide	Leukemia	Moderate
1,3-Butadiene	Leukemia	Moderate
Formaldehyde	Nasal sinuses	Moderate
	Nasopharynx	Moderate
Polychlorinated biphenyls	Liver and biliary tract	Moderate
Vinyl chloride	Liver (angiosarcoma)	Strong
	Brain	Moderate
	Lung	Moderate
	Lymphoid tissue	Moderate
Inorganic dusts and fibers		
Asbestos	Lung	Strong
	Mesothelioma	Strong
	Larynx	Moderate
	Gastrointestinal tract	Moderate
Silica, crystalline	Lung	Moderate
Talc containing asbestiform fibers	Lung	Strong
	Mesothelioma	Strong
Metals and metal compounds		
Arsenic and arsenic compounds	Skin	Strong
	Lung	Strong
	Liver (angiosarcoma)	Moderate
Beryllium	Lung	Moderate
Cadmium and cadmium compounds	Lung	Moderate
Chromium compounds, hexavalent	Lung	Strong
Nickel and nickel compounds	Nasal sinuses	Strong
	Lung	Strong
Other chemicals		
Sulfuric acid	Larynx	Moderate

Table 1 Substances that are Recognized or Probable Occupational Carcinogens, with an Indication of the Types of Cancer Affected and the Strength of the Evidence (continued)

Substance	Site	Strength of the Evidence
Radiation		
Radon	Lung	Strong
Ultraviolet	Skin	Strong
	Lip	Strong
Ionizing radiation	Leukemia	Strong
	Lung	Strong
	Breast	Strong
	Skin	Strong
	Thyroid	Strong
	Bone	Strong
	Gastrointestinal tract	Moderate
	Multiple myeloma	Moderate
	Liver	Moderate
	Bladder	Moderate
	Kidney	Moderate

ticals. This table also excludes occupations and industrial processes that have been shown to carry excess risk for workers, but where the responsible agent(s) has not been identified, thus making it difficult to use the information outside the particular setting in which the original epidemiologic study was carried out. Although the evidence concerning the substances in Table 1 was collected from studies of exposed workers, almost all these substances can be found in one form or another in the so-called general environment, in water, in air, or in food. Usually, but not always, the levels of exposure are lower and consequently the risks might be lower in the general population.

There is sometimes a tendency to interpret such tables in too categorical a fashion. The determination that a substance or circumstance is carcinogenic depends on the strength of evidence at a given point in time. The evidence is rarely clearcut but usually falls somewhere between innocence and guilt, and is liable to change as additional data accumulate. Furthermore, different carcinogens produce different levels of risk and for a given carcinogen, there may be vast differences in the risks incurred by different people exposed under different circumstances. Indeed there may be threshold effects or interactions with other factors that produce no risk for some and high risk for others.

V. INCOMPLETENESS OF CURRENT LISTS

Of the tens of thousands of chemical and physical agents that are found in the occupational environment, only a handful are considered to be human carcinogens. While this appears at first glance to be a reassuring statement, these are probably not the only occupational carcinogens. For the overwhelming majority of exposures, there is no epidemiologic evidence one way or the other concerning human carcinogenicity. Nor can it be assumed that most occupational carcinogens would have been discovered by now. There has never been a systematic and sensitive approach to discovering occupational carcinogens. Most acknowledged human carcinogens in the occupational environment were first suspected on the basis of case reports by clinicians or pathologists.[37] These discoveries were usually coincidental, requiring all of the following conditions: the occupational carcinogen affected a group of people who had some obvious common occupation or workplace; the resultant tumor was a relatively unusual one; the relative risk due to the carcinogen must have been very high for it to induce a remarkable number of cases of the rare tumor; the cluster of cases was concentrated in one medical practice; the clinician had to be very astute to notice the cluster and persistent in following it up and reporting it. The absence of any of these conditions would make it unlikely that a carcinogen would have been discovered. It is thus reasonable to suspect that those carcinogens already discovered constitute a small proportion of those actually present (viz., the tip of the iceberg). One of the foremost problems in occupational epidemiology is how to uncover the hidden part of the iceberg of occupational carcinogens. Although it is important to discover occupational carcinogens for the sake of preventing occupational cancer, the potential benefit of such discoveries goes beyond the factory walls, since most occupational exposures find their way into the general environment, sometimes at higher concentrations than in the workplace.

VI. SOURCES OF EVIDENCE ON RISK TO HUMANS DUE TO CHEMICALS

Apart from the rather anarchic, but nevertheless important process of astute clinical observation, evidence concerning carcinogenicity of a substance can come from epidemiologic studies among humans or from experimental studies of animals (usually rodents). Additionally, in the past two decades there has been increasing effort devoted to the development of so-called short-term tests, many of which are carried out *in vitro*.[38,39] Most of the short-term tests are in fact tests of mutagenicity; it is hypothesized that substances that are mutagenic may thereby be carcinogenic. A further development is in the area of studying so-called structure-activity relationships, which involves determining the carcinogenic properties of a substance on the basis of its chemical structure.[39] These latter approaches are not even close to proving their ability to predict human carcinogenic effects and will not be further discussed here.

A. EPIDEMIOLOGY

Epidemiologic research provides the most relevant data for identifying human carcinogens. Such research requires the juxtaposition of information on illness or death due to cancer in some study population and information on their past exposure to chemical and physical agents, or some indicators thereof. A third, optional data set that would improve the validity of inferences drawn from that juxtaposition is the set of concomitant risk factors that may confound the association between exposure and disease. Unfortunately, the required data are often unavailable or of questionable relevance. Because of long induction periods for cancer, it is impossible to evaluate the carcinogenicity among humans for substances that are new on the market or in the environment. Even for substances that have been with us for a long time, there are obstacles. Unlike an experimental animal for which the exposure conditions can be quite tightly controlled and monitored, each human experiences, over his or her lifetime, an idiosyncratic and bewildering pattern of exposures. Not only is it impossible to completely and accurately characterize the lifetime exposure profile of an individual, but even if we could it would be an overwhelming statistical task to tease out the effects of specific substances. The ascertainment of valid cancer diagnoses is also problematic, since subjects are often traced via routine record sources, which may be error-prone or in which cancers with long survival are poorly represented; however, this is a much less serious problem than the difficulty of determining exposure status. Since epidemiology depends on observation rather than random allocation of subjects to different exposure regimens, another problem is that the study groups being compared may very well differ in some important respects apart from exposure to the variable under study; this could distort the true association between that variable and cancer. The number of subjects available for epidemiologic study is often limited and this compromises the statistical power to detect hazards. Despite these limitations, epidemiology has made significant contributions to our knowledge of environmental carcinogens.

B. ANIMAL EXPERIMENTATION

Partly in consequence of the difficulty of generating adequate data among humans and partly because of the benefits of the experimental approach, great efforts have been devoted to studying the effects of substances in controlled animal experiments.[40] Besides the relative convenience of such an approach, it is reasonable to believe that the results generated by animal studies do have some bearing on carcinogenicity among humans and there appears to be some empiric evidence to support interspecies extrapolation.[41-43]

However, there remain serious disagreements about the inferences that can be drawn from animal experimentation.[44-50] On the one hand, the empiric evidence for a high correlation in carcinogenicity between species has been challenged,[51-53] and on the other hand, it is clear that the animal experiment is very different from the situation of free-living humans. The animal experiment is designed not to emulate the human experience but rather to maximize the sensitivity of the test to detect animal carcinogens. Doses administered to test animals are usually orders of magnitude higher than levels to which humans are exposed. The route of exposure is often unrealistic (e.g., injection or implantation) and the controlled and limited pattern of co-exposures is inevitably unlike the human situation. The "lifestyle" of the experimental animal is not only different from that of humans, but it is unlike that of its species in the wild, and for which evolution has equipped it. Animals used are typically from pure genetic strains and susceptibility to carcinogens may be higher in such populations than in genetically heterogeneous populations, such as human society. Metabolic, immunologic, and other physiologic characteristics differ between species, as do DNA repair systems. Rodents have much shorter life spans than humans and the significance of disease induction periods may differ as a result. Tumors seen in animals often occur at sites that do not have a counterpart among humans (e.g., forestomach) or which are much more rarely affected among humans (e.g., pituitary gland). Quantitative extrapolation of effects from rodents to humans depends on unverifiable mathematical assumptions concerning dose equivalents, dose-response curves, safety factors, etc. Different reasonable assumptions can lead to wildly divergent estimates.

It has been remarked, somewhat tongue in cheek, that for the purpose of identifying human carcinogens, epidemiology asks the right question but answers it badly, whereas animal toxicology asks the wrong question but answers it well. At present there is no reliable substitute for human evidence in determining risks to humans. However, epidemiologic data exist for a handful of substances, whereas experimental animal data exist for several thousand. The International Agency for Research on Cancer (IARC) recommends that in the event of sufficient evidence of carcinogenicity among animals and no evidence among humans, the substance be regarded for practical purposes as if it presents a carcinogenic risk to humans.[36] This position has come under increasing opposition from those who have questioned the relevance of animal testing for the human experience.[50]

VII. EPIDEMIOLOGIC APPROACHES TO DISCOVERING OCCUPATIONAL CARCINOGENS

The conduct of epidemiologic studies depends on idiosyncratic characteristics specific to the problem and the locale in which the study is being carried out. There are many distinct methods for studying the association between a putative risk factor and a disease. In this section, we will consider how various epidemiologic designs can be used for the purpose of investigating environmental/occupational carcinogens. The various possible approaches are distinguished primarily by the units of observation (i.e., individuals or groups of people) and by the method by which study subjects are sampled. To simplify the presentation, we will assume that individuals are either unexposed or exposed (i.e., ignore degrees of exposure).

Before discussing designs, it is useful to introduce the notion of confounding. Confounding is the phenomenon whereby a factor other than the exposure under study is associated with both the disease and the exposure factor. This will lead to a distortion in the observed association between exposure and disease. For instance, the consumption of cigarettes is higher among blue collar workers than in the rest of the population. Since smoking causes lung cancer, it follows that, even if there were no occupational carcinogens, blue collar workers would tend to have relatively high lung cancer risks. In this hypothetical example confounding would spuriously create an appearance of excess risk where there is none; confounding can work equally in the opposite direction, that is, to mask a true association. Confounding bias can result from variables that are disease risk factors, like smoking in the above example, but it can also result from differences in data collection methods or in data quality

Table 2 Layout of Cohort Data in Describing the Relationship Between a Substance and a Disease

	Total Sample[a]	Diseased	Not Diseased
Exposed	N_1	A_1	B_1
Not exposed	N_2	A_2	B_2

[a] The study sample is constituted of N_1 exposed and N_2 unexposed. The study consists of further subdividing them according to disease status.

between diseased and nondiseased subjects or between exposed and unexposed subjects.

A. COHORT STUDY

If the substance causes the disease then the risk of disease would be higher in those exposed than in those unexposed. This formulation suggests what is the most intuitively appealing form of epidemiologic study, a so-called cohort study. In a cohort study, the study sample comprises a group of people who are (or have been) exposed to the substance and another group who are (or have) not. Suppose these two groups number N_1 and N_2, respectively. Then the risk of disease in each group is estimated by tracing the study subjects to find out which ones subsequently became diseased. With an outcome such as cancer, which is relatively rare and involves a long induction period, if we were to initiate a prospective cohort study now, it might take some decades before there were enough cases of cancer to provide reasonable statistical evidence. Thus, when studying occupational circumstances, it is common to carry out the study retrospectively, that is, to define the study sample, or cohort, from a list of workers from some time in the past, and to trace them through death or incidence records up to the present. Table 2 shows schematically the layout of data generated by such a design. The risk of disease is A_1/N_1, among the exposed, and A_2/N_2 among the unexposed; the ratio of these risks is called the relative risk. A relative risk different from 1.0 indicates that there is some type of association between exposure and disease.

The fact that the cohort study approach usually involves different periods of follow-up for different subjects (i.e., workers may be hired in different years and die in different years) means that the appropriate denominator is not the number of persons but rather the number of person-years of follow-up, giving rates rather than risks. The ratio of incidence or death rates between exposed and unexposed groups is called the rate ratio. The cohort design is best suited to the study of a rare exposure that is concentrated in a small number of identifiable workplaces that are amenable to being studied. It is also best suited to the study of relatively common diseases.

Since it is often difficult or costly in practice to constitute an appropriate group of unexposed subjects with whom to compare the exposed, and since the exposed usually constitute a very small fraction of the entire population, it is expedient and often acceptable to take the disease or death rates in the entire population as a close approximation of those in the unexposed. The latter are easily available from published national statistics. When the disease experience of the exposed cohort is compared with that of the entire population, it is possible to take into account such basic demographic variables as age and sex. The most common statistical approach is a method called indirect standardization and the resulting parameter is called a standardized mortality ratio (SMR) or standardized incidence ratio (SIR). In a cohort study, it is unusual to have any additional information about subjects apart from their disease status, exposure status, age and sex. Thus, one of the main drawbacks of this design is the inability to ensure that the exposed and unexposed groups are similar in other respects. In fact, it is likely that a cohort consisting of current or ex-employees of an industry would differ in many respects from the entire national population, which includes both active and inactive persons.

A given workforce is generally exposed to a fairly narrow range of occupational substances, and for this reason the prime role of cohort studies has been and remains to investigate specific associations, rather than to explore hypotheses concerning a wide spectrum of occupations or occupational exposures. In many instances, there is little or no documented information about the past occupational exposures of individual members of the cohort. Occasionally, there are industries with long-standing hygiene measurements. However, in the few industries that have had such activities, they were generally sporadic and focused on a single agent that was already strongly suspected as harmful. Also, measurements made before 1970 (that is, the period that has considerable relevance to cancer experience today) were generally based on techniques that are of dubious validity, by today's standards.[54] The establishment of prospective disease surveillance systems by large companies is commendable and will be useful. However, such systems are recent and, because of the long latency associated with cancer induction, will only provide important results in some decades.

The cohort design has been used extensively in studies of occupational exposure, mainly because of the availability of historically valid means of reconstituting workforces from old records. There are few such opportunities for constituting cohorts of subjects who shared some general environmental exposures, such as exposure to a common drinking water source, at some time in the past.

B. CASE-CONTROL STUDY

In a case-control approach (also called case-referent), the study sample is constituted by identifying a group with the disease and another group without the disease. Data are then collected to determine how many in each group had been exposed to the factor under investigation. Within the target population of interest (e.g., city, country), the diseased sample, called cases, and the nondiseased sample, called controls, are selected to be representative of their respective groups. Suppose these number m_1 and m_2, respectively. Table 3 shows the data layout for a case-control study and shows how similar it is in form to that of a cohort study. Conceptually one may think of $a_1 + a_2$ of Table 3 to be a subsample of $A_1 + A_2$ of Table 2 and $b_1 + b_2$ of Table 3 to be a subsample of $B_1 + B_2$ of Table 2. Typically the subsample of cases would represent a large or complete sampling of $A_1 + A_2$, whereas the subsample of controls would represent a very small sampling of $B_1 + B_2$. A limitation of the case-

Table 3 Layout of Case-Control Data in Describing the Relationship Between a Substance and a Disease

	Diseased	Not Diseased
Total sample[a]	m_1	m_2
Exposed	a_1	b_1
Not exposed	a_2	b_2

[a] The study sample is constituted of m_1 cases and m_2 controls. The study consists of further subdividing them according to exposure status.

control as opposed to the cohort design is that it is difficult and often impossible to compute disease risks or rates, since the fractions $a_1/(a_1 + b_1)$ and $a_2/(a_2 + b_2)$ reflect arbitrary features of the study design rather than the disease risk. However there is a parameter, the odds ratio, defined as

$$\frac{(a_1/a_2)}{(b_1/b_2)}$$

which provides a valid estimate of the corresponding odds ratio from the cohort study.

$$\frac{(A_1/A_2)}{(B_1/B_2)}$$

If the disease is a rare one, the odds ratio is a good estimate of the relative risk of disease among those exposed. This condition holds for cancer. A difference in disease risk between exposed and unexposed is reflected in the odds ratio (OR) by a departure from a value of 1.0. Although this approach proceeds from effect to cause as opposed to the more intuitively appealing cohort study, which proceeds from cause to effect, it can be shown theoretically and it has been demonstrated empirically that it provides the same answers.[55]

As far as cancer case-control studies are concerned, the ascertainment of a case series usually poses few conceptual problems, though the practical implementation may be problematic. On the other hand the selection of an appropriate control group is often a difficult issue to resolve, the main choices usually being a set of clients selected from the same health care facilities where the cases were identified, but not having the type of cancer under study, or people selected in the general population in the same areas where the cases reside. Two principles should guide the choice of a control group. First, the controls should be representative of the persons who, if they had become ill with the disease under study, would have been included in the study as cases. Secondly, it should be possible to collect data of equivalent quality from the controls as from the cases. These two principles may lead to conflicting choices. One of the main difficulties is to ensure that cases and controls are providing data of equal quality, particularly when the data are obtained by interview with study subjects. Because of his emotional state, a cancer case may very well make a greater effort than would a nondiseased control to recall events or situations requested in the interview; on the other hand he may tend to exaggerate his exposures in a subconscious attempt to seek an explanation for his misfortune. If cases and controls provide different quality of response, this could distort the true associations.

The case-control study usually involves much more intensive interaction with study subjects than does a cohort study, often involving questionnaires and interviews. Thus, although the number of subjects included in a case-control study is typically much less than that for a cohort study, this does not necessarily translate to lower costs. In the past, most case-control studies regarding occupational cancer risks have focused on job titles held by the study subjects rather than on substances because it was thought that valid information on job exposures could not be obtained.

One of the significant benefits of the case-control approach as compared with the cohort design is the greater opportunity to obtain information concerning potentially confounding variables, which can then be taken into account in statistical analyses. The case-control approach is well suited to the situation of a relatively rare disease, such as cancer, and a relatively common exposure factor. Unlike the cohort approach, the case-control approach is quite well suited to studies of general environmental factors that were experienced long before the point in time at which the person has been ascertained for the study, but only if such factors are ones that a person can be expected to have known and be able to recall long after the event, or if records can be traced that contain some indication of the exposure, or if the exposure left some measurable biological trace in the person.

In the past, the case-control design was used relatively infrequently in occupational cancer studies, partly because of skepticism about the ability of persons to recall and accurately report details of their occupational histories. However, it has now been demonstrated that useful information can be obtained in the context of a case-control study.

First, it has been shown that interview respondents can provide accurate histories of the jobs they have held.[56-58] This could provide the basis for analyzing cancer risk in relation to job or industry titles. But this is not enough, since the purpose is to determine whether some characteristic of the work environment, usually a chemical or physical exposure, influences the risk of cancer. In this sense the job title is used as a surrogate measure of such exposures. Under certain circumstances, it may be a good surrogate and the resultant analyses would be valuable. Namely, if there is a fairly standard exposure profile among all persons who share a common job title, and if, in addition, these exposures are concentrated in a small number of job title categories, then the job title provides an efficient tool for assessing the risk due to the exposure(s). However, such a neat congruence between job titles and exposure profiles is probably the exception. Even within a company, a given job title typically covers a group of workers whose activities and chemical exposures are heterogeneous. The variation in exposures when workers are grouped by common job titles across industries and eras can be very substantial. Thus, a risk may go undetected because only a subset of a category may have been exposed to a carcinogen and the dilution may blur the association. For a product that is used by subsets of several occupational categories, considerable statistical power is lost if those workers with a common exposure cannot be pooled. If workers could be categorized on the basis of common exposure this would provide the basis for important analyses, complementing the analyses of job titles.

There are four possible approaches to infer exposure histories in the context of an interview-based case-control study: (1) environmental measurements can be carried out in study subjects' workplaces, (2) a checklist of exposures can be given to the subjects, (3) a job exposure matrix (JEM) can be applied to the job titles reported by study subjects, (4) experts can provide opinions based on whatever information is available about the subject's jobs.

Although it is conceivable to request each subject's permission to contact his employer for the purpose of carrying out environmental measurements, there are several factors that make this strategy infeasible. In general the only job in the subject's history that would be accessible to such measurement would be the current job. However, from the point of view of cancer etiology, present jobs are of less interest than past jobs; environmental measurement of a subject's current job environment is usually not indicative of the etiologically relevant exposure. The cost of measuring even one substance in each subject's current workplace would be large; the cost of measuring many substances, even if technically feasible, would be prohibitive. Finally, it is doubtful that many subjects would want the investigator to communicate with his employer and it is doubtful if employers would allow such measurements to be conducted.

A checklist of materials can be presented to subjects in the questionnaire. However, to cover the whole occupational spectrum, the checklist would have to be very long. It would be impractical to present such a long checklist without straining the subject's patience, especially if the checklist is presented for each of the subject's jobs. But of greater importance is the fact that most workers probably do not know the generic name of the "blue liquid in the green drum" or the constituents of the "smoke given off by the neighboring machine", particularly for jobs held long ago. The checklist strategy is probably quite vulnerable to a response bias, whereby cases may have a greater inclination to report exposures than do nondiseased controls.

A JEM is simply a fixed set of rules for automatically translating any job code into a list of exposures associated with the job. The application of a JEM to a set of job codes provides the basis for bringing together, for the purpose of statistical analysis, groups of workers who may share common exposures, irrespective of their occupations. Hoar et al.[59] developed a JEM for U.S. occupations. Their job axis consisted of about 500 categories, with occupations nested within industries. The exposure axis was limited to 376 substances selected because they were known or suspected carcinogens or otherwise toxic, and it was therefore expected that there would be some published documentation on the occurrence and uses of these products. In the past decade there have been a few other attempts to create JEMs, one in Britain[60] containing 50 substances, two in northern Italy,[61] of which one contained 8 recognized lung carcinogens and the other contained 74 definite or suspected carcinogens, and one other by the National Institute of Occupational Safety and Health based on national surveys of selected U.S. companies.[62]

While the use of a JEM is appealing, it is not an approach that can readily be implemented. The investigator wishing to use a JEM would either have to use one of those that has been created or develop a new one. The development of a new JEM is an enormous and usually prohibitive undertaking. None of the existing JEMs even begin to approach the size, complexity, and indeed validity of an ideal matrix. They have all been based on severely limited versions of a JEM, either in limiting the number of job categories, or the number of substances, or the validity of matrix entries, or all of these. Most have included rather broad occupational categories and rather few substances. In some of the JEMs the substances included were already known or suspected carcinogens. Such databases could not be used to discover whether any other exposures are hazardous. Occupation classification systems vary considerably from one country to another and within countries there are often more than one system in use for different purposes. A JEM based on one system is often difficult to use in another country just because of incompatibilities between job classification systems. Further, even if the job classification systems are compatible, the nature of the job exposures may differ from place to place even for the same job title. It is therefore difficult to borrow for use in one locale a JEM developed elsewhere. Even the creation of a modest-sized JEM would involve tremendous investment of resources. The validity of the entries in a matrix is necessarily limited by the resources that go into creating it: time, documentary and human consultation, and skills and experience of the persons who are responsible. The statistical benefit of using a JEM depends of course on its degree of detail and validity. The best that can be expected from a JEM is that it can bring together for analysis persons with different job titles but common exposures; however, a JEM cannot separate out subgroups of workers who, although they share a common job title, have different exposures. Only by collecting more than job and industry titles can the latter data be derived.

An alternative approach, which is probably more valid and cost-effective, involves the examination of each study subject's work history information by one or more technical experts (chemists, engineers, hygienists) for the purpose of estimating the subject's occupational exposures.[63,64] It is premised on the notion that working environments are idiosyncratic and change over time and that the consideration of each subject's job description will considerably improve the validity of exposure information. If an interview is carried out with the subject, the opportunity presents itself to obtain detailed information about each of his jobs. Based on a variety of information sources, including the subject's detailed job description, technical documentation, and consultants who are expert in one or another industry, technical experts can make educated guesses as to the possible chemical and physical exposures experienced by the subject. This is a costly approach in that it requires an ongoing presence of chemist-hygienists for the duration of the study, and there remain limitations to the validity of retrospective exposure assessment, but in terms of value for money this may be the most cost-effective approach.[65]

C. CASE-CONTROL-NESTED-WITHIN-A-COHORT STUDY

A cohort study typically involves thousands of workers traced through existing records to characterize exposure status and cancer outcome. The size of the study population usually mitigates against any attempt to obtain detailed individual data regarding exposure to the suspect agent or confounding factors. However, by focusing attention on the cases of cancer identified within the cohort and an appropriate unaffected (control) group within the cohort, it is sometimes feasible to obtain and profitably use such information. In its basic structure, this is a case-control design, but its population base is the industrial cohort rather than the general population of a geographic area. As a case-control study, it can benefit from the opportunity to collect information on disease outcome, on exposure experience, and on potential confounders related to individual study subjects. As a part of a cohort study, it can often benefit from detailed hygiene information available in company files.

D. ROUTINE RECORD STUDY OF OCCUPATION AND DISEASE OCCURRENCE

In some countries, death certificates carry information on both cause of death and occupation. This information can be exploited to compute some estimates of the association between occupation groups and risk of different causes of death. Namely, within a given occupation group, the proportion of deaths due to a given cause can be compared with the corresponding proportion among all deaths in the population, adjusted for possible differences between the populations in age at death. This so-called proportionate mortality ratio (PMR), if different from 1.0, could indicate that the risk of this cause of death is different in this occupation group than in the entire population. For instance, to evaluate the PMR for lung cancer among butchers, we would derive the proportion of all deaths that list butcher as the decedent's occupation which also record lung cancer as the cause of death and the proportion of all deaths in the entire population that record lung cancer as the cause of death; the ratio of these two proportions is the crude PMR. A sex- and age-adjusted version is easily computed if the data are available by sex and by age group. The PMR has the limitation that it can be affected not only by the cause of death that it purports to assess, but since the denominator includes all deaths in the occupation group, it is also affected by the risk of other causes of death in the occupation group. For instance, if butchers truly have a low risk of coronary heart disease compared with the entire population, then this will have the effect of inflating the butchers' PMRs for other causes of death, including lung cancer.

An alternative approach that can overcome this drawback is based on relating the number of deaths due to lung cancer among butchers not to the total number of deaths among butchers, but to the total number of butchers in the population. This number can often be estimated from the most recent national census. This allows estimation of the risk of lung cancer deaths among butchers, which can be compared with the corresponding risk of lung cancer in the entire population. (Strictly speaking these are not risks but approximations to risks.) After standardizing by age, the ratio of these two risks is a standardized mortality ratio (SMR). PMR and SMR-type indices can be computed from any routine data where disease occurrence and occupation are recorded, such as tumor registry reports. When cancer incidence, rather than cancer mortality, is used as the end-point, then the indices are called proportionate incidence ratios (PIRs) and standardized incidence ratios (SIRs).

The main advantages of this type of routine record analysis are that it can be relatively inexpensive, that the large populations covered by a national or statewide mortality or morbidity registry can provide stable disease rates, and that it can scan the occupational spectrum. Although such analyses are to be encouraged, there are many limitations, the main ones being: inability to constitute a lifelong work history of the individual, inaccuracy of job titles noted on death or tumor registry certificates, different quality of occupational information between the census and the registry, inability to obtain data on occupational exposure, inability to adjust for confounding factors apart from age and sex, errors in recorded cause of death codes, and inadequacy of cancer deaths as indicators of cancer incidence (because some types of cancer have low fatality rates). If the system is based on a tumor registry rather than death certificate reports, the last two problems mentioned would largely be avoided, since the accuracy of diagnosis is much greater in registry reports than in death certificates. But few registries routinely obtain information on occupation. Also, in some countries there is doubt about the completeness of case ascertainment in registries.

Such analyses have been carried out in several countries and states, sometimes estimating PMRs for different occupation categories,[66-68] sometimes estimating SMRs,[69] and sometimes estimating PIRs or SIRs.[70]

One of the most exciting developments in the past decade has been the increased possibility of using sophisticated record linkage methods to bring together files containing individual occupation information with those containing cancer occurrence information.[71,72] The most promising applications of this technology have occurred in Canada[73] and especially in the Nordic countries where residents have unique identifiers.[74-77] There it has been possible to link each person's census report of his occupation with his subsequent incidence of cancer in the national tumor registry, thus overcoming some of the problems with analyses based on occupations recorded on death certificates or registry reports. Further, linking the information on an individual from consecutive censuses raises the possibility of getting closer to a long-term view of a person's job history than is possible with the one-point-in-time snapshot afforded by the occupation listed on a death certificate. Not only does this approach provide better quality information on occupations and on cancer occurrence, but, depending on the type of questions asked in the census, there may be useful data available on potential confounders. Despite some problems (the sporadic nature of the occupational data, the limitation of a job title as an indicator of the risks due to occupational exposure, and the limited data available on potential confounders), the use of record linkage between censuses and tumor registries is promising.

E. GEOGRAPHIC CORRELATIONS

Unlike the other designs mentioned, which are based on the individual as the unit of observation, a geographic (or ecologic) correlation study is based on some aggregate unit, such as a county or state or country, as the unit of observation. It is premised on the notion that if an exposure causes cancer, then the areas in which the exposure is more prevalent will be those with higher cancer rates. Therefore, the correlation between the two, across areas, can be used as an indicator of the association. This is a design that is most effective when the exposure status is shared by many or all members of a given area. Thus, in some areas, most people are exposed and in other areas few are exposed. General environmental factors such as water quality and air pollution may under some circumstances fit this bill. On the other hand, occupational factors rarely do. There is rarely more than a small fraction of any given geographic area that is in a single occupation and exposed to a common set of exposures. There have been some attempts to use such an approach to study general environmental factors such as water chlorination by-products, but few attempts to use this approach to study occupational factors. Such methods suffer from several sources of bias and insensitivity.[78,79] Disease rates for the areas are diluted by the population that is not exposed. Inter-area migration further attenuates whatever effect the exposure of interest may have on inter-area cancer rates. It is difficult to obtain exposure data for the etiologically relevant era. Confounding factors are difficult, if not impossible, to control for.

VIII. CONCLUSIONS

Environmental agents play a major role in the etiology of cancer. Epidemiologic research is crucial in the investigation of risk factors for human cancer. Cancer prevention and our understanding of the process of carcinogenesis depend critically on epidemiologic research. The investigation of occupational carcinogens has been perhaps the most fruitful area for studying the role of chemical and physical agents in cancer etiology, in part because the characterization of exposure, although very problematic, is much easier in the occupational setting than in the general environment, and partly because workers have tended to be exposed to higher levels of chemicals than has the general population. Because the discovery of occupational carcinogens has depended on astute clinical observations of disease clusters rather than on systematic approaches to detection of hazards, it is likely that the list of currently recognized occupational carcinogens represents the tip of an iceberg. There are several legitimate and complementary approaches that can and should be used for the discovery and elucidation of occupational risk factors. Conventional wisdom held that hypotheses could be generated by relatively inexpensive and crude methods (e.g., geographic correlations or routine record-based SMR or PMR studies), leaving more refined methods for hypothesis-confirmation. This strategy was flawed because the crude methods were too insensitive to detect occupational cancer risks. For surveillance or hypothesis generating purposes, the most promising designs are the case-control study, with particular emphasis placed on exposure assessment, and the routine-record SMR-type study bolstered by record linkage to trace individuals. For focusing on or testing specific hypotheses, the cohort study, or its offshoot the case-control-nested-within-a-cohort study, may be the most powerful tool, although under some circumstances a population-based case-control study can also be effective.

REFERENCES

1. Hill, J., *Cautions Against the Immoderate Use of Snuff. Founded on the known qualities of the tobacco plant; and the effects it must produce when this way taken into the body: and enforced by instances of persons who have perished miserably of diseases, occasioned, or rendered incurable by its use*, 2nd ed., R. Baldwin, London, 1761.
2. Sommering, S. T., *De morbis vasorum absorbentium corporis humani*, Varrentrapp & Wenner, Frankfurt, 1795.
3. Pott, P., *Chirurgical observations relative to the cataract, polypus of the nose, the cancer of the scotum, the different kinds of ruptures and the mortification of the toes and feet*, T. J. Carnegy, London, 1775.
4. Volkmann, R., *Ueber Theer-, Paraffin und Russkrebs (Schornsteinfegerkrebs)*, in *Beitrage zur Chirurgie*, Druck und Verlag von Breitkopf und Hartel, Leipzig 1875, 370-381.
5. Bell, J., Paraffin epithelioma of the scrotum, *Edinburgh Med. J.*, 22, 135-137, 1876.
6. Morley, J., The lymphatics of the scrotum in relation to the radical operation for scrotal epithelioma, *Lancet*, 2, 1545-1547, 1911.
7. Southam, A. H. and Wilson, S. R., Cancer of the scrotum: the aetiology, clinical features, and treatment of the disease, *Br. Med. J.*, 2, 971-973, 1922.
8. Harting, F. H. and Hesse, W., Der Lungenkrebs, die Bergkrankheit in den Schneeberger Gruben, *Vierteljahresschr. Gerichtl. Med.*, 30, 296-309, 1879.
9. Rehn, L., Blasengeschwulste bei Fuchsin-Arbeitern, *Arch. Klin. Chir.*, 50, 588-600, 1895.
10. Yamagiwa, K. and Ichikawa, K., Experimental study of the pathogenesis of carcinoma, *J. Cancer Res.*, 3, 1-29, 1918.
11. Kennaway, E. L. and Hieger, I., Carcinogenic substances and their fluorescence spectra, *Br. Med. J.*, 1, 1044-1046, 1930.
12. Cook, J. W., Heiger, I., Kennaway, E. L., and Mayneord, W. V., The production of cancer by pure hydrocarbons, *Proc. R. Soc. London Ser. B*, 111, 455-484, 1932.
13. Hieger, I., The isolation of a cancer-producing hydrocarbon from coal tar, *J. Chem. Soc.*, 395, 1933.
14. Bridge, J. C., Annual Report of the Chief Inspector for the Year 1932, Her Majesty's Stationery Office, London, 1933.
15. Kuroda, S. and Kawahata, K., Uber die gewerbliche Entstehung des Lungenkrebses bei Generatorgasarbeitern, *Z. Krebsforsch.*, 45, 36-39, 1936.
16. Machle, W. and Gregorius, F., Cancer of the respiratory system in the United States chromate-producing industry, *Public Health Rep.*, 63, 1114-1127, 1948.
17. Hill, A. B. and Faning, E. L., Studies on the incidence of cancer in a factory handling inorganic compounds of arsenic. I. Mortality experience in the factory, *Br. J. Ind. Med.*, 5, 1-6, 1948.
18. Doll, R., The causes of death among gas-workers with special reference to cancer of the lung, *Br. J. Ind. Med.*, 9, 180, 1952.
19. Merewether, E. R. A., Asbestosis and carcinoma of the lung, in Annual Report of the Chief Inspector of Factories for the Year 1947, Her Majesty's Stationery Office, London, 1949, 79-81.
20. Doll, R., Mortality from lung cancer in asbestos workers, *Br. J. Ind. Med.*, 12, 81, 1955.

21. Case, R. A. M., Hosker, M. E., McDonald, D. B., and Pearson, J. T., Tumours of the urinary bladder in workmen engaged in the manufacture and use of certain dyestuff intermediates in the British chemical industry. I. The role of aniline, benzidine, alpha-naphthylamine and beta-naphthylamine, *Br. J. Ind. Med.*, 11, 75, 1954.
22. Doll, R. and Hill, A. B., Smoking and carcinoma of the lung. Preliminary report, *Br. Med. J.*, 2, 739-748, 1950.
23. Levin, M. L., Goldstein, H., and Gerhardt, P. R., Cancer and tobacco smoking, *JAMA*, 143, 336-338, 1950.
24. Mills, C. A. and Porter, M. M., Tobacco smoking habits and cancer of the mouth and respiratory system, *Cancer Res.*, 10, 539-542, 1950.
25. Schrek, R., Baker, L. A., Ballard, G. P., and Dolgoff, S., Tobacco smoking as an etiologic factor in disease. I. Cancer, *Cancer Res.*, 10, 49-58, 1950.
26. Wynder, E. L. and Graham, E. A., Tobacco smoking as a possible etiologic factor in bronchiogenic carcinoma. A study of six hundred and eighty-four proved cases, *JAMA*, 143, 329-336, 1950.
27. Doll, R. and Peto, R., *The Causes of Cancer*, Oxford University Press, Oxford, 1981.
28. Tomatis, L., Editor-in-Chief, *Cancer: Causes, Occurrence and Control*, International Agency for Research on Cancer, Lyon, 1990.
29. Franks, L. M. and Teich, N. M., *Introduction to the Cellular and Molecular Biology of Cancer*, Oxford University Press, Oxford, 1986.
30. Yuspa, S. H. and Poirier, M. C., Chemical carcinogenesis: From animal models to molecular models in one decade, in *Advances in Cancer Research*, Vol. 50, Klein, G. and Weinhouse, S., Eds., Academic Press, San Diego, 1988, 25-70.
31. Farber, E., Possible etiologic mechanisms in chemical carcinogenesis, *Environ. Health Perspect.*, 75, 65-70, 1987.
32. International Agency for Research on Cancer (IARC), *IARC Monographs on the Evaluation of Carcinogenic Risks to Humans, Vol. 14, Asbestos*, IARC, Lyon, 1977.
33. International Agency for Research on Cancer (IARC), *IARC Monographs on the Evaluation of Carcinogenic Risks to Humans, Vol. 25, Wood, Leather and Some Associated Industries*, IARC, Lyon, 1981.
34. International Agency for Research on Cancer (IARC), *IARC Monographs on the Evaluation of Carcinogenic Risks to Humans, Vol. 19, Some Monomers, Plastics and Synthetic Elastomers, and Acrolein*, IARC, Lyon, 1979.
35. International Agency for Research on Cancer (IARC), *IARC Monographs on the Evaluation of Carcinogenic Risks to Humans, Vol. 46, Diesel and Gasoline Engine Exhausts and Some Nitroarenes,* IARC, Lyon, 1989.
36. International Agency for Research on Cancer (IARC), *IARC Monographs on the Evaluation of Carcinogenic Risks to Humans, Suppl. 7, Overall Evaluations of Carcinogenicity: an Updating of IARC Monographs Vol. 1 to 42*, IARC, Lyon, 1987.
37. Doll, R., Pott and the prospects for prevention. III. 7th Walter Hubert Lecture, *Br. J. Cancer*, 32, 263-272, 1975.
38. Montesano, R., Bartsch, H., Vainio, H., Wilbourn, J., and Yamasaki, H., Eds., *Long-Term and Short-Term Assays for Carcinogens: A Critical Appraisal*, International Agency for Research on Cancer, Lyon, 1986.
39. Ashby, J. and Tennant, R. W., Chemical structure, Salmonella mutagenicity and extent of carcinogenicity as indicators of genotoxic carcinogenesis among 222 chemicals tested in rodents by US NCI/NTP, *Mutat. Res.*, 204, 17-115, 1988.

40. Rall, D. P., Hogan, M. D., Huff, J. E., Schwetz, B. A., and Tennant, R. W., Alternatives to using human experience in assessing health risks, *Annu. Rev. Public Health*, 8, 355-385, 1987.
41. Tomatis, L., Breslow, N. E., and Bartsch, H., Experimental studies in the assessment of human risk, in *Cancer Epidemiology and Prevention*, 1st ed., Schottenfeld, D. and Fraumeni, J. F., Jr., Eds., W. B. Saunders, Philadelphia, 1982, 44-73.
42. Wilbourn, J., Haroun, L., Heseltine, E., Kaldor, J., Partensky, C., and Vainio, H., Response of experimental animals to human carcinogens: an analysis based upon the IARC Monographs programme, *Carcinogenesis*, 7, 1853-1863, 1986.
43. Allen, B. C., Crump, K. S., and Shipp, A. M., Correlation between carcinogenic potency of chemicals in animals and humans, *Risk Anal.*, 8, 531-550, 1988.
44. Shubik, P., Identification of environmental carcinogens: animal test models, in *Carcinogens: Identifications and Mechanisms of Action*, Clark, G. A. and Shaw, C. R., Eds., Raven Press, New York, 1979, 37-47.
45. Kraybill, H. F., From mice to men: predictability of observations in experimental systems and their significance in man, in *Human Epidemiology and Animal Laboratory Correlations in Chemical Carcinogenesis*, Coulston, F. and Shubik, P., Eds., Ablex Publishing, Norwood, N.J., 1980, 19.
46. Roe, F. J. C., Occupational cancer: where now and where next?, *Scand. J. Work Environ. Health*, 11, 181-187, 1985.
47. Oehme, F. W., Anatomical and physiological considerations in species selections — animal comparisons, in *Human Risk Assessment — The Role of Animal Selection and Extrapolation*, Roloff, M. V., Wilson, A. G. E., Ribelin, W. E., Ridley, W. P., and Ruecker, F. A., Eds., Taylor & Francis, London 1987, 47-63.
48. Vesell, E. S., Pharmacogenetic Differences between humans and laboratory animals: implications for modelling, in *Human Risk Assessment — The Role of Animal Selection and Extrapolation*, Roloff, M. V., Wilson, A. G. E., Ribelin, W. E., Ridley, W. P., and Ruecker, F. A., Eds., Taylor & Francis, London, 1987, 229.
49. Purchase, I. F. H., *Occupational Health in the Chemical Industry*, World Health Organisation, Geneva, 1989, 102-118.
50. Ames, B. N. and Swirsky, Gold, L., Too many rodent carcinogens: mitogenesis increases mutagenesis, *Science*, 249, 970-971, 1990.
51. Purchase, I. F. H., Carcinogenic risk assessment: are animals good surrogates for man?, in *Cancer Risks: Strategies for Elimination*, Bannasch, P., Ed., Springer Verlag, Berlin, 1986, 65-79.
52. Bernstein, L., Gold, L. S., Ames, B. N., Pike, M. C., and Hoel, D. G., Some tautologous aspects of the comparison of carcinogenic potency in rats and mice, *Fundam. Appl. Toxicol.*, 5, 79-86, 1985.
53. Gold, L. S., Bernstein, L., Magaw, R., and Slone, T. H., Interspecies extrapolation in carcinogenesis: prediction between rats and mice, *Environ. Health Perspect.*, 81, 211-219, 1989.
54. Ulfvarson, U., Limitations to the use of employee exposure data on air contaminants in epidemiological studies, *Int. Arch. Occup. Environ. Health*, 52, 285-300, 1983.
55. Breslow, N. E. and Day, N. E., *Statistical Methods in Cancer Research, Vol. 1, The Analysis of Case-Control Studies*, International Agency for Research on Cancer, Lyon, 1980.

56. Baumgarten, M., Siemiatycki, J., and Gibbs, G. W., Validity of work histories obtained by interview for epidemiologic purposes, *Am. J. Epidemiol.*, 118, 583-591, 1983.
57. Bourbonnais, R., Meyer, F., and Thériault, G., Validity of self reported work history, *Br. J. Ind. Med.*, 45, 29-32, 1988.
58. Bond, G., Sobel, W., Shellenberger, R., and Flores, G., Validation of work histories obtained from interviews, *Am. J. Epidemiol.*, 122, 536-537, 1985.
59. Hoar, S. K., Morrison, A. S., Cole, P., and Silverman, D. T., An occupation and exposure linkage system for the study of occupational carcinogenesis, *J. Occup. Med.*, 22, 722-726, 1980.
60. Pannett, B., Coggon, D., and Acheson, E. D., A job-exposure matrix for use in population based studies in England and Wales, *Br. J. Ind. Med.*, 42, 777-783, 1985.
61. Macaluso, M., Vineis, P., Continenza, D., Ferrario, F., Pisani, P., and Andisio, R. Job exposure matrices: experience in Italy, in *Job Exposure Matrices*, Proceedings of a Conference held in April 1982 at the University of Southampton, Scientific Rep. No. 2, Acheson, E. D., Director, Medical Research Council, Southampton, 1983, 22-30.
62. Sieber, W. K., Sundin, D. S., Frazier, T. M., and Robinson, C. F., Development, use and availability of a job exposure matrix based on National Occupational Hazard Survey data, *Am. J. Ind. Med.*, 20, 163-174, 1991.
63. Gérin, M., Siemiatycki, J., Kemper, H., and Bégin, D., Obtaining occupational exposure histories in epidemiologic case-control studies, *J. Occup. Med.*, 27, 420-426, 1985.
64. Siemiatycki, J., Nadon, L., Lakhani, R., Bégin, D., and Gérin, M., Exposure assessment, in *Risk Factors for Cancer in the Workplace*, Siemiatycki, J., Ed., CRC Press, Boca Raton, 1991, 45-114.
65. Siemiatycki, J., Dewar, R., and Richardson, L., Costs and statistical power associated with five methods of collecting occupation exposure information for population-based case-control studies, *Am. J. Epidemiol.*, 130, 1236-1246, 1989.
66. Guralnick, L., Mortality by occupation and cause of death among men 20-54 years of age: United States, 1950, (Vital Stat. Special Reports, Vol. 43, No. 5), U.S. Department of Health, Education, and Welfare, Washington, D.C., 1963.
67. Milham, S. J., *Occupational Mortality in Washington State, 1950-1971*, DHEW Pub. No. (NIOSH) 76-175-A, B, C,, U.S. Government Printing Office, Washington, D.C., 1976.
68. Dubrow, R. and Wegman, D. H., Cancer and occupation in Massachusetts: a death certificate study, *Am. J. Ind. Med.*, 6, 207-230, 1984.
69. Office of Population Censuses and Surveys, Occupational mortality, The Registrar General's decennial supplement for England and Wales, 1970—72, Series DS no. 1, Her Majesty's Stationary Office, London, 1978.
70. Menck, H. R. and Henderson, B. E., Occupational differences in rates of lung cancer, *J. Occup. Med.*, 18, 797-801, 1976.
71. Howe, G. R. and Spasoff, R. A., *Proceedings of the Workshop on Computerized Record Linkage in Health Research*, University of Toronto Press, Toronto, 1986.
72. Baldwin, J. A., Acheson, E. D., and Graham, W. J., Eds., *Textbook of Medical Record Linkage*, Oxford University Press, New York, 1987.

73. Howe, G. R. and Lindsay, J., A generalized record linkage computer system for use in medical follow-up studies, *Comput. Biomed. Res.*, 14, 327-340, 1981.
74. Lynge, E. and Thygesen, L., Occupational cancer in Denmark. Cancer incidence in the 1970-census population, *Scand. J. Work Environ. Health*, 16, 1-35, 1990.
75. Borgan, J. K. and Kristofersen, I. B., *Mortality by Occupation and Socio-Economic Group in Norway 1970-1980*, Statistisk Sentralbyra, Oslo, 1986.
76. Pukkala, E., Occupation, socioeconomic status and education as risk determinants of cancer, *Cancer Detect. Prev.*, 14, 1989.
77. Statistiska Centralbyran, *Death Register 1961-70*, Statistiska Centralbyran, Stockholm, 1982.
78. Siemiatycki, J., Day, N. E., Fabry, J., and Cooper, J. A., Discovering carcinogens in the occupational environment: a novel epidemiologic approach, *J. Natl. Cancer Inst.*, 66, 217-225, 1981.
79. Morgenstern, H., Uses of ecologic analysis in epidemiologic research, *Am. J. Public Health*, 72, 1336-1344, 1982.

CHAPTER 8

The Epidemiology of Waterborne Disease: The Importance of Drinking Water Disinfection

Gunther F. Craun

I. INTRODUCTION

Waterborne diseases are transmitted through the ingestion of contaminated water; water acts as the passive carrier of the agent. The occurrence of waterborne outbreaks and epidemics has been well documented. The endemic transmission of waterborne diarrhea and other infectious diseases has recently been recognized by environmental epidemiologic studies.

In the U.S., waterborne diseases have largely been controlled through the use of multiple barriers of protection: wastewater treatment, selection of best available water sources, protection of source water quality, and treatment of drinking water by disinfection and filtration. Outbreaks occur when one or more of these systems fail. In less developed countries where all of these multiple barriers may not be utilized, the waterborne transmission of infectious diseases is still a major problem, causing both mortality and morbidity.

Water disinfection is an important part of the multiple barrier concept. Its role in preventing waterborne disease should not be overlooked because of recent concerns about possible cancer risks of disinfection by-products. Any potential risks that may be associated with drinking water disinfection must be balanced with the benefits provided by disinfection. This chapter considers the epidemiology of waterborne disease and discusses these benefits and potential risks.

A. ACUTE ILLNESS

Inhalation and dermal exposures to water contaminants are not considered in this chapter, but they may be important sources of infection for some waterborne patho-

0-87371-573-X/95/$0.00+$.50

gens.[1-4] *Pseudomonas* folliculitis has been caused by water contact in hot tubs and whirlpools; pharyngitis, external otitis, and schistosome dermatitis have been associated with recreational activities in swimming pools and lakes. *Legionella* can also contaminate drinking water systems, find conditions favorable for growth, and may be aerosolized into the indoor and outdoor air environment. Pontiac fever caused by *Legionella* has been associated with use of whirlpools, and legionellosis has been associated with contamination of the outdoor environment through cooling towers. Nontuberculosis mycobacteria can also contaminate water systems and may cause infection in individuals with immune deficiencies.

Also not considered in this chapter are the water-based and water-vectored diseases. Water-based diseases where the pathogen spends an essential part of its life in water or is dependent upon aquatic organisms for completion of its life cycle include schistosomiasis, dracontiasis, and primary amoebic meningoencephalitis. Yellow fever, dengue, filariasis, malaria, onchocerciasis, and sleeping sickness are water-vectored diseases transmitted by insects that breed in water or bite near water.

Depending upon the etiologic agent and host response, symptoms of waterborne disease include asymptomatic infection, mild diarrhea, and more severe reactions which may result in death.[5-11] Among the etiologic agents responsible for waterborne illness throughout the world, some pathogens are important in the U.S. and other industrialized countries while others are not.

The ease with which people can travel from country to country has provided additional opportunities for the transmission of many waterborne diseases in the developed countries. Pathogens may be carried by refugees and immigrants settling in the U.S. and by citizens returning from travel throughout the world. Whether a pathogen becomes an important cause of waterborne illness in the U.S. depends upon many factors, including the infective dose, excretion patterns of the organism, environmental conditions that allow survival of the pathogen, level of protection against contamination of water sources, and the adequacy of treatment of water supplies to remove or inactivate the pathogen. When only a small number of organisms are required for infection, pathogens can easily reach infective dose levels in surface water and groundwater sources if these sources are not protected from sewage discharges.

Chemical contamination of drinking water can also result in illness. Chemical poisonings are among the most commonly recognized causes of waterborne outbreaks in the U.S. and commonly occur from cross-connections in the water distribution system and the local contamination of groundwater sources. Acute waterborne chemical poisonings usually result in mild, self-limited illness,[6] but six deaths from chemical contamination have been reported in the U.S. since 1971. The symptoms of chemical poisonings vary with the substance ingested. Both fluoride and heavy metals cause vomiting within 1 h after consumption of the contaminated water. Infantile methemoglobinemia associated with high nitrate levels can result in death if not medically treated.[2]

Environmental epidemiologic studies have suggested associations between low-level chronic exposure to several chemical contaminants in water and a variety of adverse health effects.[8,12-19] Inorganic water contaminants of importance include lead,

Table 1 Historically Important Waterborne Diseases

Disease	Incubation Period
Bacterial	
Shigellosis	1–7 d
Salmonellosis	6–72 h
Typhoid Fever	1–3 d
Cholera	1–5 d
Protozoan	
Giardiasis	7–14 d
Amebiasis	14–48 d
Viral	
Hepatitis A	15–45 d

Table 2 Recently Identified Waterborne Pathogens

Pathogen	Incubation Period
Bacterial	
Enterotoxigenic E. *coli*	12–72 h
Enterohemorrhagic E. *coli*	12–60 h
Campylobacter fetus spp. *jejuni*	1–7 d
Yersinia enterocolitica	1–7 d
Protozoan	
Cryptosporidium	2–10 d
Viral	
Norwalk agent and Norwalk–like virus	12–48 h
Rotavirus	2–3 d

nitrate, selenium, arsenic, fluoride, and sodium. Over 1000 organic compounds have been identified in drinking waters and their potential human health risks are currently being evaluated by toxicologic and epidemiologic studies. Inhalation is an important route of exposure for the volatile organic water contaminants, whereas ingestion is likely more important for inorganic chemicals and organic compounds that are not volatile.[20,21] The inhalation route of exposure is also important for radionuclides. For example, the off-gassing of radon from naturally contaminated groundwater has been identified in some areas as a significant source of environmental exposure.[22]

II. INFECTIOUS WATERBORNE DISEASE

A. WATERBORNE PATHOGENS

A number of pathogens have been identified as causes of waterborne disease[23] in the U.S. (Tables 1 and 2). Cholera, the classic waterborne disease, is internationally reported.[7,24] Infected persons may have no symptoms, mild to moderate diarrhea, or severe watery diarrhea, vomiting, and dehydration. Without medical treatment, dehydration often leads to death in severe cases. *Vibrio cholerae* 01 is categorized into two serotypes (Inaba and Ogawa) and two biotypes (Classical and El Tor). The infectious dose is large, requiring some 100,000 to 10,000,000 organisms to produce infection. Since 1961 the seventh cholera pandemic, caused by the El Tor biotype, has spread from Asia to the Middle East, Oceana, Africa, and the Americas.[7] Epidemic cholera, which had not been reported in South America since 1895, was introduced to Peru in January 1991 and quickly spread to other South American countries and Mexico. Contaminated drinking water is an important source of infection. Tourists to South America have returned with infections, but no sustained transmission has occurred in the U.S. An epidemic focus of cholera, however, does exist in the U.S, and since 1973, 65 cases have been reported, primarily associated with consumption of undercooked crabs or shrimp or raw oysters from the Gulf of Mexico. A waterborne outbreak occurred in 1981 off the Texas coast; *V. cholera* 01 was responsible for 17 cases of severe diarrhea caused by sewage contamination of an oil rig's potable water system through a cross-connection.

Two enteric diseases commonly transmitted by drinking water are *Shigella* and *Salmonella* gastroenteritis. The symptoms, diarrhea, abdominal cramps, fever, and vomiting, are similar, but *Shigella* infection produces bloody stools and rectal pain more frequently.[7] A sudden onset of headache is usually reported with salmonellosis. Humans are the only significant reservoir of infection for *Shigella*. For *Salmonella*, domestic and wild animals can also be important sources of infection. Infection with *Shigella* requires only a few organisms, from 10 to 100; usually more than 100 organisms is required for *Salmonella*.[7]

Typhoid fever (*Salmonella typhi*) is an enteric infection often transmitted by drinking water in past years. Patients usually experience fever, headache, and abdominal pain, and the diagnosis is confirmed by blood, urine, or stool culture.[7] Contaminated water is also responsible for outbreaks of paratyphoid fever, which has similar symptoms but tends to be milder. Humans are important reservoirs of infection for both diseases, but dairy cattle and other domestic animals may be important reservoirs of infection for paratyphoid fever. Although few cases of typhoid fever are reported in the U.S., waterborne outbreaks occasionally occur. An outbreak of 210 cases was reported in 1973 after contamination of an untreated well serving a migrant labor camp in Florida;[25] 60 cases resulted from the contamination of water mains in a housing development in 1985 in St. Croix, Virgin Islands.[26]

The protozoan *Giardia lamblia* is a frequent cause of waterborne disease outbreaks in the U.S. Although it can produce a long-lasting diarrhea accompanied by fatigue, weight loss, epigastric pain, bloating, and flatulence,[7,27] a large number of infections are asymptomatic. Outbreaks are frequently caused by the ingestion of untreated stream water in mountainous areas and public water supplies that use surface waters without filtration. *Giardia* cysts can be inactivated by chlorine, but high concentrations and/or long contact times are required; water filtration with appropriate pretreatment can remove cysts from water.[5] The infective dose is felt to be low, with less than 100 organisms required for infection. Both humans and animals are sources of infection. *Giardia* spp. have been identified in mammals, birds, and amphibians, and a variety of wild and domestic mammals have been shown to become infected in laboratory studies when exposed to *Giardia* cysts from humans.[28] Beavers have been implicated in several outbreaks in surface water systems, but other animals may also be important sources of water contamination.

Sometimes transmitted by contaminated drinking water, *Entamoeba histolytica* and *Balantidium coli* are not important causes of waterborne diseases in the U.S. These protozoan cysts are also resistant to the usual disinfection dosage used in drinking water, but only five waterborne outbreaks involving 75 cases of amebiasis have been reported in the U.S. since 1946. Pigs are the principal reservoir of *B. coli*, and this infection is associated primarily with poor sanitation conditions. For example, an outbreak of 110 cases of balantidiasis occurred in Micronesia in 1971 when a typhoon destroyed water catchment facilities and people were forced to utilize water supplies contaminated by pig feces.[29]

Hepatitis infections are caused by several viral agents and differ in etiology, clinical signs, and epidemiology.[7] Many infections are asymptomatic or cause a mild illness, but a severely disabling disease can result. Symptoms of fever, nausea, vomiting, muscle aches, and jaundice occur approximately 15 to 45 d after exposure. The diagnosis is made by history, physical examinations, tests of liver function, and serologic studies that demonstrate IgM antibodies against the virus. Hepatitis A virus, long recognized to be transmitted by drinking water, has recently been identified in water samples collected during outbreaks.[30] Viral hepatitis E (non-A, non-B hepatitis) can also be transmitted by contaminated water, and waterborne outbreaks have been reported outside the U.S.[7] Viral hepatitis B, C, and D are transmitted through contaminated blood and other body fluids by transfusions, sexual activities, perinatal exposure, accidental needle sticks, or use of contaminated needles among drug users.[7]

As significant advances in laboratory techniques have occurred, more viral agents are detected through immunoassay or electron microscopy. Several viruses have been identified in outbreaks of waterborne gastroenteritis.[5,32] Acute infectious nonbacterial gastroenteritis is caused by at least two distinct groups of viruses. The 70-nm rotaviruses are a major cause of acute gastroenteritis in infants and young children, typically causing severe diarrhea for 5 to 8 d, frequently accompanied by fever and vomiting.[7] The illness is usually sporadic, but is occasionally epidemic, and although it occurs predominantly in young children, it can affect adults. Rotavirus infection may be responsible for half the cases of infant diarrhea that require hospitalization worldwide and an estimated 870,000 deaths each year in less developed countries.[7]

The other viruses associated with acute nonbacterial gastroenteritis are less well characterized, and at least three serologically distinct 27-nm virus-like particles have been identified. The Norwalk agent is the prototype and most extensively studied of this group. Gastroenteritis is typically an explosive but self-limited illness, usually lasting 24 to 48 h and characteristically epidemic. Symptoms include vomiting, diarrhea, nausea, abdominal cramps, headache, low-grade fever, anorexia, and malaise.[7] The 27-nm viruses are responsible for family and community outbreaks among older children and adults and are reported to cause one third of the epidemics of viral gastroenteritis reported in the U.S.[7]

Viral agents are suspected as a major cause of waterborne gastroenteritis in outbreaks where an etiology is not determined, but bacteria and protozoa may also be important causes of these outbreaks.[23] Enterotoxigenic and enterohemorrhagic *Escherichia coli*, *Campylobacter fetus* spp. *jejuni*, *Yersinia*, and *Cryptosporidium* have been identified as waterborne etiologic agents during recent years.[1,2,5,23,31] *E. coli* and other coliform bacteria are indicators of water contamination. Although these bacteria are generally not pathogenic, enterotoxigenic, enteroinvasive, enteropathogenic, and enterohemorrhagic strains of *E. coli* have been identified.[7] The enteroaggregative strain causes infant diarrhea in less developed countries, but is not well defined. Humans are reservoirs of infection for all strains of *E. coli*, but cattle are believed to be an important reservoir of infection for enterohemorrhagic *E. coli*. Enteroinvasive strains cause disease manifested by fever and mucoid and occasionally bloody diarrhea. Enterotoxigenic strains cause disease ranging from mild diarrhea to seriously dehydrating cholera-like illness with profuse watery diarrhea, abdominal cramps, and vomiting; they are a common cause of travelers' diarrhea. The enteropathogenic strains have generally been associated with outbreaks of acute diarrheal disease in newborn nurseries and epidemics of watery diarrhea with mucus, fever, and dehydration in infants less than 1 year of age. The waterborne transmission of enterohemorrhagic *E. coli* 0157:H7 has recently been documented.[1] In 1989 an outbreak of 243 cases occurred in Cabool, Missouri, where one third of the ill persons had bloody diarrhea, 32 were hospitalized, and 4 died. The water distribution system was identified as the likely source of contamination. The severity of illness associated with this organism reinforces the need to prevent waterborne outbreaks.

Y. enterocolitica is frequently found in surface and well waters and frequently isolated from wild and domestic animals, the principal reservoir of infection.[23] Fever, abdominal pain, headache, and watery diarrhea are the most common symptoms of infection. The number of cases that are recognized is small and largely dependent on the experience of the clinician and laboratory. The first well-documented waterborne outbreak in the U.S. in 1981 was caused by the contamination of a spring used as a water source for packaging tofu.[33] Although *Yersinia* was isolated from well water implicated in a gastroenteritis outbreak at a ski resort in Montana in 1974, clinical specimens were not examined for this organism, and investigators did not attribute the illness to this organism.[34] A number of individual cases of suspected waterborne transmission have been reported where similar serotypes are isolated from the patient and water, but these have not been reported as outbreaks.

There are several species of *Campylobacter*, but the major pathogen of humans is *C. jejuni*, an important cause of bacterial gastroenteritis and travelers' diarrhea.[7,23] Most infections in developed countries are seen in young adults; in less developed countries, most infections occur in children. Illness is frequently self-limiting, lasting no more than 1 week, and mild infections may produce symptoms resembling those seen in viral gastroenteritis. However, *C. jejuni* may also cause a typhoid-like illness, and some patients have a prolonged illness or relapse. *Campylobacter* species are widespread in animals and may be transmitted from animal reservoirs to humans by direct contact with infected animals or through the ingestion of contaminated food or water. Water contamination is frequently caused by wild and domestic animals. The infective dose is greater than 500 organisms.[7] The first waterborne outbreak of gastroenteritis caused by *C. jejuni* in the U.S. occurred in 1978, affecting approximately 3000 people in Bennington, Vermont.[35]

Cryptosporidium, an intestinal protozoan parasite, is a well-known cause of diarrhea in animals, but has been recognized only recently as a cause of human disease.[7,10,11,36,37] Like *Giardia*, animals as well as humans may serve as sources of environmental contamination and human infection. The first case of human cryptosporidiosis was reported in 1976.[38,39] Some 58 cases were reported during 1976 to 1983; 18 patients with normal immune function had self-limited diarrhea, but 40 patients with immunologic abnormalities developed severe, irreversible diarrhea, causing death in 22 patients. The first reported waterborne outbreak of cryptosporidiosis in July 1984 was traced to sewage contamination of an artesian well used by the community of Braun Station, Texas.[36] Untreated well water was found to be contaminated with fecal coliforms, the well water was chlorinated, and tap water samples collected prior to the outbreak were negative for coliforms.

In 1987, *Cryptosporidium* caused an estimated 13,000 cases of illness in Carrollton, Georgia, a community that used surface water that underwent conventional treatment — coagulant feed, rapid mix, flocculation, sedimentation, filtration through anthracite sand filters, and chlorination.[37] Water quality regulations were not exceeded for turbidity or coliforms, and disinfection was not interrupted. *Cryptosporidium* oocysts identified in the water system during the outbreak were felt to be associated with turbidity breakthrough or passage of floc through rapid sand filters, which had been taken out of service for a period of time and restarted without backwashing. Waterborne outbreaks of cryptosporidiosis have been reported in both filtered and unfiltered water supplies in the United Kingdom[40] and Oregon.[41] More recently in 1993 a waterborne outbreak of over 350,000 cases of illness was traced to *Cryptosporidium* contamination of drinking water in Milwaukee, Wisconsin.[42] This is the largest waterborne outbreak reported in the U.S. since record keeping began in 1920 and occurred even though drinking water regulations were not exceeded.

Limited studies have been conducted, but evidence indicates *Cryptosporidium* is highly resistant to water chlorination.[5,10,11,31,43] Only 90% inactivation of oocysts was found at 100 mg/l free chlorine for 2 h contact, whereas 99% inactivation of *Giardia* occurs at about 2.5 h at 1 mg/l, polio virus at 3 min at 1 mg/l, rotavirus and *E. coli* bacteria at less than 1 minute at 0.1 mg/l.* The occurrence of *Cryptosporidium* is widespread in water supplies in the U.S.[44] Oocysts were detected in 55% of 257 samples collected from surface water and springs in 17 states and in 17% of 36 tap water samples. In comparison, *Giardia* cysts were found in 16% of the same water samples, and none were found in the tap water samples. Since water analyses cannot assess viability, additional research is required to determine the waterborne risk of cryptosporidiosis. Cases may not be well recognized and reported, asymptomatic infection may be high, and protective immunity may be important.

A protozoan pathogen similar to *Cryptosporidium* in morphologic features and the diarrheal disease it causes was recently identified.[45] Previously referred to as

* Information for comparative purposes only; EPA regulations require inactivation of 99.9% for *Giardia* and 99.99% for viruses.

cyanobacterium-like or coccidian-like bodies (CLB), the name *Cyclospora* sp. has been proposed for this newly recognized pathogen. CLB have been identified worldwide in stool specimens from patients with diarrhea, and this may be the same pathogen identified in a 1990 waterborne outbreak[1] in a Chicago hospital. CLB but no ova or parasites were found in stool specimens from patients who had relapses of explosive watery diarrhea lasting up to 4 weeks. Open-air, roof-top storage tanks were used to maintain water pressure; no CLB were found in water samples but algae were found in the storage tank. In Nepal, an epidemiologic study of diarrheal patients with CLB identified in stool specimens found water to be one route of transmission of the infection; CLB was also identified in the implicated water source.[46] The complete life cycle, pathogenic mechanism, and epidemiology of this newly identified waterborne etiologic agent remain to be fully described.

Almost half of the reported waterborne outbreaks in the U.S. have resulted in acute gastroenteritis whose etiology is not determined. Inadequate investigation or insufficient laboratory analyses may be responsible for failing to detect an etiologic agent in many of these outbreaks. However, in some outbreaks extensive laboratory analyses failed to identify an etiology. For example, a causative agent has yet to be identified for a distinctive chronic diarrheal illness characterized by dramatic, urgent, watery diarrhea persisting for many months. The first outbreak of this illness was reported in Brainerd, Minnesota, in 1984 after consumption of unpasteurized milk.[47] In 1987 the first waterborne outbreak of chronic gastroenteritis was reported; untreated, contaminated well water in an Illinois restaurant was implicated as the vehicle of transmission.[48] Diarrhea with a median frequency of 12 stools per day persisted in 87% of patients after 6 months. No bacterial, mycobacterial, viral, or parasitic agents known to be enteropathogenic were detected in stools or well water. A second outbreak also associated with untreated well water occurred in a small Oklahoma community in 1988, but again no agent could be isolated.[1] Although more timely and complete investigations may help identify the more common etiologic agents, increased availability of specialized laboratory procedures and additional research are also needed to identify newly recognized or suspected agents.

B. ENDEMIC WATERBORNE DISEASE

Several recent environmental epidemiologic studies have found an increased risk of giardiasis and cryptosporidiosis in unfiltered and untreated water supplies in the U.S.; this risk was seen among endemic cases rather than epidemic cases and suggests that the waterborne disease risk in industrialized countries may be greater than previously realized.[49-55] Most studies of endemic waterborne disease found an increased risk associated with use of untreated surface water, especially when camping or hiking. In New Mexico, an increased risk of cryptosporidiosis was associated with the use of untreated surface water.[55] Use of shallow well sources was found to be an additional risk factor for giardiasis in New Hampshire; 18% of the endemic giardiasis was attributed to use of shallow wells or surface water sources by individuals for drinking water. In Vermont, a higher incidence of nonepidemic giardiasis was found in municipalities using unfiltered surface water or wells and in areas with private water systems compared to municipalities with filtered surface waters. Although the drinking water in Dunedin, New Zealand, is chlorinated, the incidence of laboratory-confirmed giardiasis was found to be threefold higher in the area of the city where surface water is not filtered.[56]

The risk of endemic waterborne illness has also been described in a filtered municipal water system near Montreal, Canada, which had not exceeded current water quality limits. Some 35% of the unreported gastrointestinal illness experienced by consumers of tap water was attributed to waterborne transmission and considered preventable.[57,58] The water source is a sewage-contaminated river that is conventionally treated with predisinfection, alum coagulation, flocculation, rapid sand filtration, ozone, and chlorine. A randomized intervention trial showed a 50% higher incidence of self-reported mild, gastrointestinal illness among those individuals who drank tap water over a 15-month period compared to individuals in households supplied with reverse osmosis water filters at their tap to remove microbial and chemical contaminants.[57,58] Increased bacterial growth on the reverse-osmosis units was also found during the study, and heterotrophic bacterial counts (organisms grown at 35°C) in water from these filters may also have been associated with gastrointestinal illness.[57] The risk associated with the use of contaminated filters, however, was felt to be very small compared to the risk attributed to tap water.

C. WATERBORNE OUTBREAK STATISTICS IN THE U.S.

Since 1920, 1768 waterborne outbreaks with 472,288 cases of illness and 1091 deaths have been reported in the U.S. Almost all deaths were due to typhoid fever prior to 1940, but 12 deaths have occurred since 1971. Four deaths were caused by *E. coli* 0157:H7 and two by *Shigella*. The remaining deaths were due to chemicals — arsenic, nitrate, fluoride, ethylene glycol. A total of 345 waterborne outbreaks were reported during 1981 to 1992 in community (38%) and noncommunity (36%) systems and from the ingestion of contaminated water from recreational (17%) and individual (9%) water sources. An average of 29 waterborne outbreaks occurred each year during this period, slightly less than reported during the previous decade and comparable to the number reported during the 1930s and 1940s.[59] Most outbreaks occurred in public water systems: small communities and areas with noncommunity systems; contaminated, untreated groundwater, or inadequately disinfected groundwater was responsible for 43% of all reported waterborne outbreaks (Table 3). Contaminated, untreated, or inadequately treated surface water was responsible for 22% of all reported outbreaks.[53]

In surface water systems, outbreaks occurred primarily because of inadequate or interrupted disinfection in systems that did not provide filtration. A large number of outbreaks was reported in filtered surface water systems, and this emphasizes the importance of operation and monitoring if water filtration is to be an effective treatment barrier against the transmission of waterborne disease.

In community water systems, most of the outbreaks occurred in unfiltered surface water sources (27%) or were caused by distribution system contamination (24%), primarily through cross-connections and repairs of mains (Table 3). Almost all

Table 3 Causes of Waterborne Outbreaks in the U.S., 1981–1992

Cause of Outbreak	Type of Water System[a] Community	Noncommunity	Other
Inadequate disinfection of surface water when only treatment	34	10	—
Contamination of distribution system	32	7	1
Filtration deficiencies			
Surface water	20	1	—
Groundwater	2	—	—
Untreated groundwater	16	51	21
Inadequate disinfection of groundwater when only treatment	16	43	1
Unknown, miscellaneous	7	8	4
Inadequate chemical feed	3	—	—
Untreated surface water	2	4	4
Inadequate treatment for chemical removal	—	—	1
Ingestion during water recreation	—	—	57
Totals	132	124	89

[a] Values represent number of outbreaks.

Table 4 Etiology of Waterborne Outbreaks in the U.S., 1981–1992

Disease	Number of Outbreaks	Cases of Illness
Acute gastroenteritis (AGI)	155	42,163
Giardiasis	79	7,136
Shigellosis	28	4,416
Acute chemical poisoning	21	578
Viral gastroenteritis	17	9,944
Hepatitis A	12	338
Campylobacterosis	10	1,436
Yersiniosis	2	103
AGI (*E. coli* 0157:H7)	2	323
Chronic Gastroenteritis	2	94
Leptospirosis	2	14
Typhoid fever	1	60
Cyanobacterium-like bodies	1	21
Cholera	1	17
Amebiasis	1	4
Totals	345	85,061

outbreaks (76%) in noncommunity systems were caused by contaminated, untreated, and inadequately disinfected groundwater. Although groundwater was generally found to be contaminated by sewage overflows and septic tank effluents, wild or domestic animals were also possible sources of contamination, either from surface runoff into wells or springs or the influence of streams on groundwater.

The most frequently identified waterborne etiology during 1981 to 1992 was giardiasis (Table 4). Almost all waterborne giardiasis outbreaks reported in the U.S. since 1965 were caused by contamination of surface water (67%) and sewage contamination of groundwater or groundwater under the influence of surface water (12%). While most of the giardiasis outbreaks occurred in surface water systems where disinfection was the only treatment, ineffective filtration or pretreatment of surface water was also responsible for outbreaks.

D. STRATEGIES FOR PREVENTION OF WATERBORNE DISEASE

Waterborne outbreaks and endemic waterborne disease have occurred in areas where water systems have not exceeded regulations for coliforms and turbidity, and disinfection has not been effective as the only treatment for surface waters. New strategies are required to prevent the waterborne transmission of infectious disease because of threats from newly recognized pathogens. Historically, the important waterborne diseases were typhoid fever and cholera — diseases that are transmitted exclusively among humans. Sources of infection for recently identified waterborne pathogens may include wild and domestic animals in addition to humans. Protection of water sources from human sewage contamination and chlorination introduced early in this century were successful in preventing the waterborne transmission of cholera and typhoid fever. *V. cholerae* and *S. typhi* are very susceptible to chlorine disinfection and require a relatively large number of organisms to cause infection. Waterborne outbreaks are now caused by *Giardia* and *Cryptosporidium*, protozoa that are much more resistant to chlorine and require few organisms to cause infection. Water disinfection may be sufficient to inactivate coliform bacteria without inactivating the protozoa, and water supplies that meet the coliform limits may still pose a risk if water sources are not sufficiently protected from contamination and are inadequately treated. Removal of *Cryptosporidium* from drinking water is critical because of its resistance to chlorine.

The protection of source water quality from discharges of human waste continues to be important, but is only one barrier for prevention of waterborne disease. Since animals may be important sources of water contamination for these protozoa, greater emphasis must be placed on appropriately designed and operated water treatment barriers. Except for high quality water sources, both filtration and disinfection treatment barriers are required to provide adequate margins of safety for surface water systems. The Environmental Protection Agency (EPA) surface water treatment requirements (54 FR 27486-541, June 19, 1989) address disinfection and filtration of surface waters in the U.S. for *Giardia*; a revised surface water treatment rule is under consideration for *Cryptosporidium*.

EPA has also recently drafted requirements for disinfection of public water systems using groundwater in the U.S. Adequate disinfection of groundwater should reduce the occurrence of waterborne outbreaks and endemic waterborne disease, especially in small systems where intermittent contamination of wells and springs cannot be readily determined or prevented. It is important to remember that the multiple barrier approach applies to groundwater as well as surface water sources. Disinfection is not a substitute for the proper development and protection of groundwater sources from contamination such as septic tanks, cesspools, sewerage lines, surface runoff. Disinfection of groundwater is an additional barrier which can provide an increased level of protection.

III. POTENTIAL HEALTH RISKS OF WATER DISINFECTION

A. OCCURRENCE OF DISINFECTION BY-PRODUCTS

The chlorination of drinking water, when properly applied as part of the multiple barrier system, has been successful in preventing the transmission of infectious waterborne diseases, but undesirable chlorination by-products may be formed in some water systems. In 1974 it was found that chlorine can react with naturally occurring humic and fulvic substances in drinking water sources to produce trihalomethanes (THMs), a group of halogen-substituted single carbon compounds.[60,61] The predominate THMs are chloroform and bromodichloromethane, and their concentration is dependent on the presence and concentration of the precursor substances, the species, dosage, and contact time of chlorine, and the water pH and temperature.[62] Levels of THMs are generally higher when surface water is chlorinated than when groundwater is chlorinated.[63] Lower levels of THMs are found in waters disinfected with chlorine and ammonia, but this depends upon the specific treatment process.[64] If chlorine is added before ammonia, chlorine will have an opportunity to react with available precursors to form by-products. If ammonia is added first or at the same time as chlorine, the formation of chlorinated by-products should be minimal.

Halogenated compounds in addition to THMs[19,65-67] can also be formed during the chlorination of drinking water (Table 5). A survey[68,69] of water treatment facilities in the U.S. showed total THMs to be the most frequently detected by-product with a median value of 39 μg/l. The haloacetic acids with a median value of 19 μg/l were the next most significant by-product. Median values for aldehydes and haloacetonitriles ranged from 2 to 4 μg/l, and haloketones ranged from 1 to 2 μg/l. Cyanogen chloride was found to be preferentially produced when chloramine is used as a disinfectant. Formaldehyde and acetaldehyde, primarily by-products of ozone, were also detected in chlorinated water supplies.

Water pH affects the formation of the halogenated by-products.[19] During water treatment at low pH, the formation potential of THMs is minimized, but the formation of most other chlorinated organics may be maximized. Lowering the pH to reduce THMs may also increase the formation of corrosion by-products and any associated health risks. Chlorinated by-products can be reduced by moving the point of chlorination after coagulation-flocculation and controlling precursor substances through physical removal mechanisms, but if the water contains bromide, brominated by-products will be formed.[18,67] By-products can be removed after their formation, but the required treatment technology (activated carbon, air stripping, membranes) may be too complex and not economically feasible for many communities. Using alternate disinfectants such as chloramine, ozone, or chlorine dioxide is also an option, but these alternate disinfectants may produce other disinfection by-products of potential concern. Selecting a water disinfectant requires consideration not only of its ability to form potentially toxic by-products but also its ability to inactivate

Table 5 Partial List of Byproducts of Chlorine and Ozone Disinfection of Drinking Water

Byproducts
Chlorine:
Chloral hydrate
Chlorinated hydroxyfuranones
Chloropicrin
Chlorophenols
Cyanogen chloride
Haloacetonitriles
Haloacetic acids
Haloaldehydes
Haloketones
Trihalomethanes
Ozone:
Organic
Acids
Alcohols
Aldehydes
Epoxides
Ketones
Nitrosamines
Peroxides
Quinones
Inorganics
Hydrogen peroxide
Bromate
Iodate
Chlorate

Adapted from Clark, R. M., *Health Environ. Digest*, 4(3), 4, 1990.

waterborne pathogens. Water disinfected with chloramine, for example, may contain low levels of THMs, but chloramine is a less effective disinfectant than free chlorine; longer contact times and higher concentrations are required to inactivate waterborne pathogens. Chloramine can cause acute effects in kidney dialysis patients, such as methemoglobinemia, hemolysis, and anemia, and to avoid these complications, dialysis centers must provide additional treatment to remove monochloramine.[70] Chlorine dioxide, an effective water disinfectant, does not produce THMs, but other by-products are produced, including chlorite and chlorate, which may cause acute health effects at low concentrations. THMs are not a by-product of ozone, an oxidant and disinfectant used widely in Europe and becoming more common in the U.S. However, other by-products are produced, and little is known about their occurrence and possible health effects.

While ozone is an effective disinfectant, no disinfectant residual can be maintained in the water distribution system. A disinfectant residual is important protection against possible contamination and bacterial regrowth in the water distribution system. The evaluation of potential risks of a disinfectant must consider the effects of any residual disinfectant concentration as well as the by-products that may be produced. Carcinogenesis has been the primary concern for by-products; toxicologic effects other than carcinogenesis are important when considering disinfectant residuals.[18]

B. MUTAGENICITY

Increased levels of mutagenic activity, as determined by the *Salmonella*/microsome assay, are usually present in chlorinated drinking water compared to the corresponding untreated water.[19,71] Studies suggest that the reaction of chlorine with natural humic substances is the likely source of mutagen formation in drinking water and that water chlorination is responsible for the majority of the increased mutagenic activity.[19,71-74]

Results of studies assessing the mutagenicity of water disinfectants have been fairly consistent for chlorine, chloramine, and chlorine dioxide. The largest amount of mutagenicity is produced by chlorine disinfection; chloramine and chlorine dioxide usually produce less mutagenicity. Ozone treatment of water has been found to both increase and decrease mutagenic activity; in some studies mutagenic activity was increased to intermediate levels or levels as high as obtained with chlorination. Mutagenicity can, at least partly, be removed by coagulation-flocculation and filtration. Granular activated carbon treatment is also effective.

3-Chloro-4-(dichloromethyl)-5-hydroxy-2(5H)-furanone (MX) and its isomer E-2-chloro-3-(dichloromethyl)4-oxobutenoic acid (E-MX) have been detected in humic acid solutions and chlorinated drinking waters.[19,74] The concentration of MX found in chlorinated drinking water ranged from 5 to 67 ng/l. MX is a potent mutagen and may be responsible for up to 57% of mutagenicity in chlorinated drinking water; E-MX is less mutagenic.

Mutagenic testing may be of little value in assessing the carcinogenic risks of disinfectants and disinfection by-products. A quantitative relationship may not exist between mutagenic as currently measured and carcinogenic risks, activity, and mutagenic potency may be unable to predict carcinogenic potency.[18] The *Salmonella*/microsome assay has also been shown to have poor sensitivity and specificity in predicting carcinogenic activity; only 58% of 115 known animal carcinogens were detected in the assay and only 70% of 107 noncarcinogens tested as negative.[18] No excess tumors were attributed to chlorinated humic acids when provided to B6C3F1 mice for 24 months, even though solutions of chlorinated humic acids displayed mutagenic activity.[18] Although mutagenic activity was detected in chlorination by-products from Mississippi River water by the *Salmonella*/microsome assay, these same by-products were unable to produce detectable carcinogenic activity in the mouse skin initiation/promotion assay, strain A lung adenoma assay, or rat liver foci assay.[18]

C. TOXICOLOGY

Chlorine, hypochlorous acid, or hypochlorite have not been shown to have carcinogenic properties.[18] Although adverse effects on the cardiovascular system were suggested by studies of pigeons consuming chlorinated water, more recent studies in rabbits, micropigs, and rats did not show that lipid metabolism was affected by ingestion of chlorine.[19,75] As previously noted, the regulation of chlorine will likely be based on the potential carcinogenicity of its by-products rather than the

Table 6 Estimated Carcinogenic Risks from Toxicologic Studies*

By–product	Chlorine	Chloramine	Ozone
Chloroform	0.24	0.48	0
Bromodichloromethane	0.91	0.18	
Chlorodibromomethane	0.29	0.06	
Bromoform	0.05	0.01	0.03
Dichloroacetic acid	<0.01	<0.01	
Trichloroacetic acid	54	11	
Chloropicrin	<0.01	<0.01	
2,4,6–trichlorophenol	0.02	<0.01	
Formaldehyde	0.54	0.11	3.0
Hydrogen peroxide			10
Bromate			50
Mean risk (MLE)	56	11.8	63

* Additional cases per million people from drinking for a lifetime. The mean concentration of the indicated by-product found in U.S. drinking water systems.

Adapted from Bull, R.J. and Kopfler, F. C., *Health Effects of Disinfectants and Disinfection By-Products*, American Water Works Association, Denver, CO, 1991.

potential toxicity of residual levels of chlorine.[18] For example, limits for residual levels of chlorine in drinking water, if based on toxicologic studies showing weight loss and production of sperm head abnormalities, would be well above the concentrations commonly used in water treatment.

Maximum likelihood estimates (Table 6) of carcinogenic risks, based on published toxicologic and water quality information, have been determined for individual by-products of several disinfectants.[18] These estimates of excess risk utilize the multistage model to calculate the probability that an individual during his lifetime would develop cancer from water containing the mean concentration of the specified by-product as found in U.S. drinking water systems. A mean overall risk, expressed in additional cases per million people per lifetime, is obtained by summing the risks for individual disinfectant by-products.

THMs have been shown to induce tumors in one or more species or strains of experimental animals.[18] However, the mechanism by which the THMs produce cancer has not been established. In mice, tumors were observed in the liver and kidney; in rats, tumors were seen in the kidney, thyroid, and intestine. The mathematical model for estimating cancer risk assumes that at any dose there is some probability of inducing cancer and that the chemical acts as an initiator. Studies, however, suggest that chloroform administered in corn oil may act as a promotor of spontaneously initiated cells in the liver of mice rather than as a tumor initiator.

The by-product that largely affects the estimated cancer risk is trichloroacetic acid (TCA). This by-product appears to produce liver tumors in B6C3F1 mice but has not yet been shown to cause tumors at any site in F344 rats.[18] The relevance of liver tumors observed in B6C3F1 mice for predicting cancer in humans is controversial. The liver of mice is very sensitive to toxins, and many compounds cause liver tumors in mice but do not cause tumors in rats or other species.[76] In addition, spontaneous tumor formation rates of 30% have been observed in B6C3F1 mice.[77] When the data from toxicologic studies of liver tumors in B6C3F1 mice are included in the assessment of risk, an additional 56 cancers per million people are estimated for chlorinated by-products, but when these data are excluded, only 1.8 additional cancers are estimated.

The toxicology of chronic exposure to chloramine is not well established, and the carcinogenicity of chloramine is currently being evaluated.[18] Current estimates of carcinogenic risks associated with chloramine by-products are based on the assumption that by-products similar to chlorinated by-products would be formed at 20% of the levels observed with chlorination. Actual levels of by-products are dependent on the manner in which chlorine and ammonia are reacted to form chloramine. An exception is the by-product cyanogen chloride, which has been associated with chloramine but not chlorine.

Evidence of altered brain development in two strains of rat appears to be the most critical effect of chlorine dioxide and suggests a drinking water limit for residual chlorine dioxide of 0.1 to 0.2 mg/l.[18] Chlorite is the major by-product and hemolytic anemia is the major adverse health effect. Chlorate, also a by-product of chlorine dioxide disinfection, has been used as a weed killer, and the current toxicologic evidence is insufficient to confidently assure the safety of chlorine dioxide as a disinfectant. Anticipated limits for residual levels of chlorine dioxide, chlorate, and chlorite are such that it is unlikely effective disinfection with chlorine dioxide can be achieved.

As used in drinking water, ozone is not expected to produce residuals at the tap, and health concerns are limited to its by-products. Ozone is frequently used in combination with hydrogen peroxide; this combination increases the amount of hydroxyl radicals. By-products include a variety of aldehydes, ketones, carboxylic acids, and peroxides (Table 5). Hydrogen peroxide, formaldehyde, and bromate have been found to induce cancer in experimental animals; other by-products have not been tested. Bromate in drinking water has been shown to increase kidney tumors in male and female rats and appears to be the major by-product affecting risk. Water systems with bromide in source waters may contain bromate as a by-product of the use of ozone.

The strength of toxicologic evidence for carcinogenicity varies among the disinfection by-products, and it should be remembered that many by-products have not been adequately tested. Also of concern is the use of the multistage model to provide a standard comparison. Since the cancer mechanisms are not known, other mathematical models might be more appropriate and could result in different estimates of risk. Thus, the risks presented in Table 6 must be considered preliminary and for comparative purposes only. Based on current toxicologic information, there appears to be very little difference in the overall carcinogenic risk associated with the use of chlorine or ozone. However, the specific by-products responsible for the greatest risk differ between chlorine and ozone; TCA is the most important by-product for chlorine and bromate the most important for ozone.

D. EPIDEMIOLOGY

In the 1970s, descriptive epidemiologic studies in the U.S. suggested increased risks of cancer of the bladder, stomach, large intestine, and rectum in areas with chlorinated surface waters.[14,19,78-83] These associations encouraged analytical epidemiologic studies which, unlike descriptive studies, consider an individual's disease status, exposure to chlorinated water, and potential confounding and modifying exposures or risk factors. All but one of the analytical studies were case-control design. In case-control studies where the only information used was obtained from death certificates, investigators found bladder mortality associated with chlorinated drinking water in two of five states; colon and rectal cancer mortality were associated with chlorinated drinking water in three states.[84-88] However, these studies are of limited value because potential confounding bias and nonrandom misclassification of exposure are likely to result from incomplete information available from the death certificates on residential histories, exposure to chlorinated water, and possible confounders.

In several traditional case-control studies where more complete information was obtained by interviews, a small increased cancer risk was reported[89-93] among people who used chlorinated drinking water for a very long time, but in other studies no increased cancer risk was found.[94-97] Chlorinated drinking water was found to raise the risk of colon cancer among the elderly in a case-control study in North Carolina,[92] but was not associated with colon cancer in another study in Wisconsin.[93] In Wisconsin and New York exposure to chlorinated by-products was estimated, and colon cancer was not found to be associated with THMs.[95,96]

A case-control study[89-91] of bladder cancer incidence in five states and five metropolitan areas of the U.S. reported an increased bladder cancer risk, but only among a small, otherwise low-risk population of nonsmokers who drank chlorinated water for over 60 years. Further analysis of this same study indicated bladder cancer risk was also associated with high water consumption among long-term users of chlorinated surface water, but inconsistencies were noted among males and females.[19] Analysis of data from the Iowa portion of this national study showed an overall increased risk of bladder cancer among study participants who had been exposed to chlorinated water, and a large case-control study is currently being conducted in Iowa to further assess the risk of bladder, colon, and rectal cancer.[19,93]

In Massachusetts, a case-control study of populations using the same surface water source disinfected with chlorine or chloramine found two to three times higher bladder cancer mortality in the population receiving chlorinated water.[98-100] Because water disinfected with chloramine contains lower THMs levels, this suggests the higher bladder cancer mortality in chlorinated water supplies might be associated with THMs. In the Iowa analysis, bladder cancer risks were found to be higher among residents receiving water that had been chlorinated before filtration, a practice that is likely to result in high THMs.[93] These epidemiologic studies suggest, but provide no direct evidence, that THMs may be associated with a cancer risk. No association was observed between estimated levels of THMs and colon cancer, and it is possible that historical exposures to THMs were not accurately estimated or were too low to detect an association in these studies. Other water contaminants, such as synthetic volatile organics in chlorinated municipal groundwater, the large number of organic contaminants in surface waters, or other chlorinated by-products, might also be confounding the epidemiologic results.

The current epidemiologic evidence is limited and, at best, only suggestive of an association between chlorinated drinking water and an increased risk of cancer.[19,101] There is always uncertainty in epidemiologic studies because of the limitations of studying human populations and assessing individual exposures. To assess cancer risks associated with chlorinated water systems, information is available from only a few analytical epidemiologic studies. The interpretation of associations observed for rectal and bladder cancer in the five death certificate studies is limited because of possible systematic bias. Five additional case-control studies and one cohort study obtained information about residential histories and confounding factors from sources other than death certificates. The cohort study is of limited value; the risk estimates are statistically unstable because of the small study population. Only one of three interview case-control studies found an association with colon cancer, and this association was observed only in the elderly. An increased risk of bladder cancer was observed in two interview case-control studies, but the risks were observed only after very long exposures to chlorinated water. The evidence from epidemiologic studies for an association between chlorinated water and increased cancer is limited, and the information is insufficient to suggest a casual association.

Establishing causality from epidemiologic studies requires sufficient evidence of a dose-response, consistent results from several studies which do not have major methodological limitations and bias, a moderately strong association as judged by the relative risk, and a plausible biological mechanism of disease. After causality is established, the potential magnitude of the cancer risk that may be attributed to the exposure of interest should be determined. It is important to estimate these risks because the disinfection of drinking water provides significant benefits. Actions taken to decrease potential cancer risks may increase waterborne infectious disease risks, and the risks and benefits must be compared.

The International Agency for Research on Cancer (IARC) in 1991 reviewed the toxicologic studies designed primarily to investigate the potential effects of organic extracts of drinking water and epidemiologic studies of chlorinated water exposures. The IARC concluded[102] that "Chlorinated drinking-water (was) *not classifiable as to its carcinogenicity to humans*", because of "*inadequate evidence* for the carcinogenicity of chlorinated drinking-water in humans" or "in experimental animals". Additional epidemiologic studies reported since this evaluation do not provide sufficient information to alter these conclusions: (1) a descriptive study in Norway that found an ecologic association between water chlorination and colorectal cancer but cautioned that no evidence of a causal relationship was provided;[103] (2) incomplete information from a case-control study in Colorado that suggested an increased risk of bladder cancer in populations with more than 30 years exposure to chlorinated surface water but no increased risk with levels of THMs or residual chlorine;[104] (3) a case-control study in Washington County, Maryland, that reported an increased risk of pancreatic cancer associated with chlorinated municipal water but cautioned about limitations in the assessment of water exposures.[105]

The same epidemiologic studies evaluated by the IARC were statistically combined in a meta-analysis to obtain pooled estimates of relative and attributable risk.[106,107] The meta-analysis reported an association between chlorination by-products in drinking water and bladder and rectal, but not colon, cancer with pooled relative risk estimates of 1.15 for bladder cancer and 1.38 for rectal cancer. While meta-analysis has been used for clinical studies, it may not be appropriate for observational epidemiologic studies.[107] In evaluating the results of a meta-analysis, two important points must be remembered: (1) systematic biases or methodological problems of individual epidemiologic studies used in the analysis are not corrected by the analysis, and (2) meta-analysis provides no additional information on possible causality of the observed associations. The overall results of the analysis, especially for rectal cancer, may be unduly affected by studies with major methodological limitations, and no sensitivity analyses were conducted to determine how the results were affected by including the death certificate case-control studies, which suffer from confounding and misclassification biases.

The relatively small estimate of risk reported in the meta-analysis does not offer evidence of a strong epidemiologic association between chlorinated water and cancer risk. In epidemiologic studies, the magnitude of the relative risk estimate can be used to help judge the strength of the association and whether the observed association might be spurious. Uncontrolled confounding bias is less likely to influence an epidemiologic association when the relative risk is large (greater than 1.5). It is difficult to interpret a weak association (a relative risk of less than 1.5) because one or more weak confounding characteristics can lead to such an association and it is usually not possible to identify and measure all of the weak confounding characteristics.

The inconsistent results from the various epidemiologic studies may be due to inaccurate estimates of exposure from both misclassification[108] and poor techniques for their assessment. Water consumption was estimated in only one study, and inhalation exposures from the volatile by-products have not been considered. In most of the analytical epidemiologic studies, cancer risks were compared in populations receiving chlorinated water and populations receiving unchlorinated water. Because few surface waters are unchlorinated, the studies essentially compared chlorinated surface water exposures with unchlorinated groundwater exposures. Thus, exposure to other surface water contaminants may be contributing to the observed cancer risks. The majority of epidemiologic studies considered chlorinated water supplies, and studies are needed in areas where water has been disinfected with ozone for a long period of time. Studies are also needed to confirm the lower bladder cancer mortality observed in populations where chloramine was used as a disinfectant.

Associations between chlorinated water and cardiovascular effects have been reported, but these are much less certain than cancer associations. One epidemiologic study[109,110] reported an association between serum cholesterol and chlorinated drinking water, but another study[111] did not find an association. Several small clinical trials have found no evidence that drinking water containing either chlorine or monochloramine affects lipid metabolism.[112-116]

Recent epidemiologic studies[117-120] have suggested that water chlorination and certain by-products of water chlorination may be associated with adverse reproductive outcomes, including low birthweight and congenital malformations. However, the studies contain a number of biases and methodological limitations, and the results must be interpreted as preliminary and suggestive. Additional studies are required to determine if adverse reproductive outcomes are epidemiologically associated with water chlorination.

IV. CONCLUSIONS

Disinfection is required to protect against the transmission of waterborne disease. Concerns about the cancer risks that may be associated with long-term exposures to chlorine or chlorinated by-products must be tempered with considerations of the benefits provided by water chlorination. In many developing countries, water chlorination is an economical, practical technology and is important in preventing a significant number of illnesses and deaths from infectious waterborne diseases. Alternate disinfectants may be too expensive and not technologically feasible in these areas. Almost 60% of the waterborne outbreaks reported during 1981 to 1992 in the U.S. occurred as the result of inadequate or no chlorination of surface and groundwater, and this underscores the need for disinfection even in the highly developed countries where waterborne infectious diseases are largely under control. Endemic waterborne disease risks reported in the industrialized countries also indicate that microbial risks may be more important than previously realized. A reminder of the continuing problem of microbial risks is provided by the large waterborne outbreak of cryptosporidiosis in Milwaukee in 1993.

Although questions about the safety of water chlorination still remain, much more is known now than in 1976 when THM limits were established because of concerns in the U.S. about the carcinogenicity of chloroform in experimental animals. Many other by-products have recently been identified as the result of disinfection with chlorine or alternate disinfectants. Mutagenic tests have not been useful in quantitative risk assessments but have helped identify water contaminants for toxicity testing. Toxicologic studies have identified carcinogens among the disinfection by-products, but most toxicologic studies show hepatic tumors in mice, whereas the epidemiologic studies suggest weak associations between chlorinated water and bladder and possibly rectal cancer. Inadequate toxicologic data and uncertainties about the carcinogenic risk estimate models limit the conclusions that can be reached about the disinfectant that may pose the lowest carcinogenic risk.

There is also uncertainty about the epidemiologic information. Agreement has not been reached among epidemiologists that chlorinated drinking water causes cancer. However, the question is often asked — what is the risk attributable to chlorinated drinking water if the reported cancer association were causal? An estimate of the magnitude of the cancer risk is usually made after a causal association has been established according to accepted epidemiologic principals. Morris et al.,[106] however, have estimated the bladder and rectal cancer risks attributable to chlorinated surface water in the U.S. They suggest that 9% or 4200 cases of bladder cancer and 15% or 6500 cases of rectal cancer per year may be associated with the consumption of chlorinated surface water. These attributable risk estimates must be interpreted very cautiously, however. The estimates are based on limited information, which includes studies with major methodological problems, and a high degree of uncertainty is associated with these estimates of risk.[121]

There seems little reason from the toxicologic evidence to abandon chlorine as a water disinfectant; alternate disinfectants do not appear to offer significant advantages. Chloramine is not a very effective disinfectant. Chlorine dioxide cannot be used unless techniques are provided to reduce residual levels of chlorine dioxide, chlorate, and chlorite. The estimated risks for ozone by-products are similar to chlorinated by-products. A lifetime risk of an additional 56 cases of cancer per million people was estimated for chlorine by-products from the toxicologic studies; 63 additional cases were estimated for ozone by-products. Chlorine will continue to be a widely used water disinfectant.

Although more research is needed to assess the potential health risks of disinfection by-products, prudent public health practice suggests that exposures should be reduced when possible. However, changes in disinfection practices to reduce exposures to by-products must not increase infectious disease risks. It must be remembered that disinfection is the final barrier against transmission of waterborne pathogens, not the only barrier.[59] Selection of the best available water source and protection of the source water quality is important for both surface water and groundwater. For all but the exceptional high quality surface water sources, filtration and disinfection will be required to prevent the transmission of waterborne disease. Properly designed and operated filtration plants make disinfection more effective by reducing microbiological contamination and removing turbidity and other substances that exert chlorine demand and interfere with the efficacy of the disinfection process. When disinfection is a part of the multiple barrier approach, precursor concentrations may be reduced and lower concentrations of the disinfectant can be used, thereby lowering the levels of by-products produced. Disinfection treatment must be properly designed to ensure that it is continuous and effective in decreasing infectious waterborne disease risks.

REFERENCES

1. Herwaldt, B.L., Craun, G.F., Stokes, S.L., and Juranek, D.D., Waterborne-disease outbreaks, 1989-1990, *Morbid. Mortal. Weekly Rep.*, 40 (SS-3), 1, 1991.
2. Levine, W.C. and Craun, G.F., Waterborne-disease outbreaks, 1986-1988, *Morbid. Mortal. Weekly Rep.*, 39 (SS-1), 1, 1990.
3. Dufour, A.P., Diseases caused by water contact, in *Waterborne Diseases in the United States*, Craun, G.F., Ed., CRC Press, Boca Raton, FL, 1986, 23.
4. Thornsberry C., Balows, A., Feeley, J.C., and Jakubowski, W., Eds., *Legionella*, American Society for Microbiology, Washington, D.C., 1984.

5. Craun, G.F., Ed., *Methods for the Investigation and Prevention of Waterborne-Disease Outbreaks*, Environmental Protection Agency, Cincinnati, OH, 1990.
6. Harris, J.R., Clinical and epidemiological characteristics of common infectious diseases and chemical poisonings caused by ingestion of contaminated drinking water, in *Waterborne Diseases in the United States*, Craun, G.F., Ed., CRC Press, Boca Raton, FL, 1986, 11.
7. Benenson, A.S., Ed., *Control of Communicable Diseases in Man*, 15th ed., American Public Health Association, Washington, D.C., 1990.
8. Committee on Drinking Water, *Drinking Water and Health*, Vol. 1, National Academy of Sciences, Washington, D.C., 1977.
9. Craun, G.F., Waterborne giardiasis, in *Giardiasis*, Meyer, E.A., Ed., Elsevier, Amsterdam, 1990, 267.
10. Sykora J.L. and Craun, G.F., Eds., *The Taxonomy, Detection, Epidemiologic, and Waterborne Control of Cryptosporidium*, University of Pittsburgh, Pittsburgh, PA, 1989.
11. Report of the group of experts, Badenoch, J., Chairman, *Cryptosporidium in Water Supplies*, Her Majesty's Stationery Office, London, 1990.
12. Craun, G.F., Chemical drinking water contaminants and disease, in *Waterborne Diseases in the United States*, Craun, G.F., Ed., CRC Press, Boca Raton, FL, 1986, 43.
13. Craun, G.F., Health aspects of groundwater pollution, in *Groundwater Pollution Microbiology*, Bitton, G. and Gerba, C.P., Eds., John Wiley & Sons, New York, 1984, 135.
14. Committee on Drinking Water, *Drinking Water and Health*, Vol. 3, National Academy of Sciences, Washington, D.C., 1980.
15. Committee on Drinking Water, *Drinking Water and Health*, Vol. 4, National Academy of Sciences, Washington, D.C., 1982.
16. Committee on Drinking Water, *Drinking Water and Health*, Vol. 5, National Academy of Sciences, Washington, D.C., 1983.
17. Committee on Drinking Water, *Drinking Water and Health*, Vol. 6, National Academy of Sciences, Washington, D.C., 1986.
18. Bull, R.J. and Kopfler, F.C., *Health Effects of Disinfectants and Disinfection By-Products*, American Water Works Association, Denver, CO, 1991.
19. Craun, G.F., Epidemiologic studies of organic micropollutants in drinking water, in *The Handbook of Environmental Chemistry, Vol 5, Pt. A, Water Pollution*, Hutzinger, O., Ed., Berlin, Springer-Verlag, 1991, 1.
20. Wallace, L.A., Comparison of risks from outdoor and indoor exposure to toxic chemicals, *Environ. Health Perspect.*, 95, 7, 1991.
21. Andelman, J.B., Human exposures to volatile halogenated organic chemicals in indoor and outdoor air, *Environ. Health Perspect.*, 62, 313, 1985.
22. Hess, C.T., Weiffenbach, C.V., and Norton, S.A., Environmental radon and cancer correlations in Maine, *Health Phys.*, 45, 339, 1983.
23. Craun, G.F., Statistics of waterborne outbreaks in the U.S. (1920-1980), in *Waterborne Diseases in the United States*, Craun, G.F., Ed., CRC Press, Boca Raton, FL, 1986, 73.
24. Craun, G., Swerdlow, D., Tauxe, R. et al., Prevention of cholera in the United States, *J. Am. Water Works Assoc.*, 83, 40, 1991.
25. Saslow, M.S., Nitzkin, J.L., Feldman, R. et al., Typhoid fever; public health aspects, *Am. J. Public Health*, 65, 1184, 1975.
26. St. Louis, M.E., Water-related disease outbreaks, 1985, *Morbid. Mortal. Weekly Rep.*, 37 (SS-2), 15, 1988.

27. Meyer, E.A., Ed., *Giardiasis*, Elsevier, Amsterdam, 1990.
28. Craun, G.F. and Jakubowski, W., Status of waterborne giardiasis outbreaks and monitoring methods, in *Proceedings of the International Symposium on Water Related Health Issues*, Tate, C.L., Jr., Ed., American Water Resources Association, Bethesda, MD, 167, 1986.
29. Center for Disease Control, *Morbid. Mortal. Weekly Rep.*, 21, 59, 1972.
30. Sobsey, M.D., Fuji, T., and Hall, R.M., Inactivation of cell-associated and dispersed hepatitis A virus in water, *J. Am. Water Works Assoc.*, 83, 64, 1991.
31. Craun, G.F., Surface water supplies and health, *J. Am. Water Works Assoc.*, 80, 240, 1988.
32. Craun, G.F., Ed., *Waterborne Disease Outbreaks, Selected Reprints of Articles on Epidemiologic, Surveillance, Investigation, and Laboratory Analysis*, Environmental Protection Agency, Cincinnati, OH, 1990.
33. Centers for Disease Control, Outbreak of *Yersinia enterocolitica* — Washington State, *Morbid. Mortal. Weekly Rep.*, 31, 562, 1982.
34. Eden, K.V., Rosenburg, M.L., Stoopler, M. et al., Waterborne gastroenteritis at a ski resort — isolation of *Yersinia enterocolitica* from drinking water, *Public Health Rep.*, 92, 245, 1977.
35. Vogt, R.L., Sours, H.E., Barrett, T. et al., Campylobacter enteritis associated with contaminated water, *Ann. Intern. Med.*, 96, 292, 1982.
36. D'Antonio, R.G., Winn, R.E., Taylor, J.P. et al., A waterborne outbreak of cryptosporidiosis in normal hosts, *Ann. Intern. Med.*, 103, 886, 1985.
37. Hayes, E.B., Matte, T.D., O'Brien, T.R. et al., Large community outbreak of cryptosporidiosis due to contamination of a filtered public water supply, *N. Engl. J. Med.*, 320, 1372, 1989.
38. Current, W.L., *Cryptosporidium*: its biology and potential for environmental transmission, *CRC Crit. Rev. Environ. Control*, 17, 21, 1986.
39. Fayer, R. and Ungar, L.P., *Cryptosporidium* spp. and cryptosporidiosis, *Microbiol. Rev.*, 50, 458, 1986.
40. Report of the group of experts, Badenoch, J., Chairman, *Cryptosporidium in Water Supplies*, Her Majesty's Stationery Office, London, 1990.
41. A Large Outbreak of Cryptosporidiosis in Jackson County, Communicable Disease Summary, Oregon Health Division, 1992, 41.
42. MacKengie, W.R., Hoxie, N.J., Proctor, M.E. et al. A massive outbreak in Milwaukee of *Cryptosporidium* infection transmitted through the public water supply, *N. Engl. J. Med.*, 331, 161-7, 1994.
43. Korick, D.G., Mead, J.R., Madore, M.S., Sinclair, N.A., and Sterling, C.S., Effects of ozone, chlorine dioxide, chlorine, and monochloramine on *Cryptosporidium parvum* oocysts viability, *Appl. Environ. Microbiol.*, 56 (5), 1423, 1990.
44. Rose, J.B., Gerba, C.P., and Jakubowski, W., Survey of potable water supplies for *Cryptosporidium* and *Giardia*, *Environ. Sci. Technol.*, 25, 1393, 1991.
45. Ortega, Y.R., Sterling, C.R., Gilman, R.H., Cama, V.A., and Diaz, F., Cyclospora species — a new protozoan pathogen of humans, *N. Engl. J. Med.*, 18, 1308, 1993.
46. Hoge, C.W., Shlim, D.R., Rajah, R., Triplett, J. et al., Epidemiologic of diarrheal illness associated with coccidian-like organism among travellers and foreign residents of Nepal, *Lancet*, 341 (8854), 1175, 1993.
47. Osterholm, M.T. et al., An outbreak of a newly recognized chronic diarrhea syndrome associated with raw milk consumption, *JAMA*, 256, 484, 1986.

48. Parsonnet, J., Trock, S.C., Bopp, C.A. et al., Chronic diarrhea associated with drinking untreated water, *Ann. Intern. Med.*, 110, 985, 1989.
49. Wright, R.A., Spenser, H.C., Brodsky, R.E., and Vernon, T.M., Giardiasis in Colorado: an epidemiologic study, *Am. J. Epidemiol.*, 105, 330, 1977.
50. Weiss, H.B., Winegar, D.A., Levy, B.S., and Washburn, J.W., Giardiasis in Minnesota, 1971-1975, *Minn. Med.*, 60, 815, 1977.
51. Harter, L., Frost, F., and Jakubowski, W., *Giardia* prevalence among 1 to 3 year old children in two Washington state counties, *Am. J. Public Health*, 72, 386, 1982.
52. Chute, C.G., Smith, R.P., and Baron, J.A., Risk factors for endemic giardiasis, *Am. J. Public Health*, 77, 585, 1987.
53. Birkhead, G. and Vogt, R.L., Epidemiologic surveillance for endemic *Giardia lamblia* infection in Vermont, *Am. J. Epidemiol.*, 129, 762, 1989.
54. Laxter, M.A., Potential exposure of Utah Army National Guard personnel to giardiasis during field training exercises: a preliminary survey, *Mil. Med.*, 150, 23, 1985.
55. Gallaher, M.M., Herndon, J.L., Nims, L.J. et al., Cryptosporidiosis and surface water, *Am. J. Public Health*, 79, 39, 1989.
56. Fraser, G.G. and Cooke, K.R., Endemic giardiasis and municipal water supply, *Am. J. Public Health*, 81, 760, 1991.
57. Payment, P., Richardson, L., Siemiatycki, J., Dewar, R., Edwardes, M., and Franco, E., A randomized trial to evaluate the risk of gastrointestinal disease due to consumption of drinking water meeting current microbiological standards, *Am. J. Public Health*, 81, 703, 1991.
58. Payment, P., Franco, E., Richardson, L., and Siemiatycki, J., Gastrointestinal health effects associated with the consumption of drinking water produced by point of use domestic reverse osmosis filtration units, *Appl. Environ. Microbiol.*, 57, 945, 1991.
59. Craun, G.F., Waterborne disease outbreaks in the United States of America: causes and prevention, *World Health Stat. Q.*, 45, 192, 1992.
60. Bellar, T.A., The occurrence of organohalides in chlorinated drinking water, *J. Am. Water Works Assoc.*, 66, 703, 1974.
61. Rook, S.S., Formation of haloforms during chlorination of natural waters, *Water Treatment Exam.*, 23, 234, 1974.
62. Symons, J.M., Stevens, A.A., Clark, R.M. et al., *Treatment Techniques for Controlling Trihalomethanes in Drinking Water*, Environmental Protection Agency, Cincinnati, OH, 1981.
63. Burke, T.A., Amsel, J., and Cantor, K.P., Trihalomethane variation in public drinking water supplies, in *Water Chlorination: Environmental Impact and Health Effects*, Vol. 4, Jolley, R.L., Brungs, W.A., Cotruvo, J.A. et al., Eds., Ann Arbor Science Publishers, Ann Arbor, MI, 1983, 1343.
64. Rice, R.G. and Gomez-Taylor, M., Occurrence of by-products of strong oxidants reacting with drinking water contaminants: scope of the problem, *Environ. Health Perspect.*, 69, 31, 1986.
65. Johnson, J.D., Christman, R.F., Norwood, D.C. et al., Reaction products of aquatic humic substances with chlorine, *Environ. Health Perspect.*, 46, 63, 1982.
66. Oliver, B.G., Dihaloacetonitriles in drinking water: algae and fulvic acid as prercursors, *Environ. Sci. Technol.*, 17, 80, 1983.
67. Clark, R.M., Controlling disinfection by-products: a research challenge, *Health Environ. Digest*, 4 (3), 4, 1990.
68. Krasner, S.W., McGuire, M.J., Jacangelo, J.G. et al., The occurrence of disinfection by-products in U.S. drinking water, *J. Am. Water Works Assoc.*, 81, 41, 1989.

69. McGuire, M.J. and Meadow, R.G., AWWARF trihalomethane survey, *J. Am. Water Works Assoc.*, 80, 61, 1988.
70. Craun, G.F., Drinking water disinfection: assessing health risks, *Health Environ. Digest*, 4 (3), 1, 1990.
71. Noot, D.K., Anderson, W.B., Daignault, S.A., Williams, D.T., and Huck, P.M., A review of the use of the Ames mutagenicity assay to evaluate drinking water treatment processes, *J. Am. Water Works Assoc.*, 81, 87, 1989.
72. Coleman, W.E., Munch, J.W., Kaylor, W.H. et al., Gas chromatography/mass spectrometry of mutagenic extracts of aqueous chlorinated humic acid — a comparison of the by-products to drinking water contaminants, *Environ. Sci. Technol.*, 18, 674, 1984.
73. Meier, J.R., Ringhand, H.P., Coleman, W.E., Schenck, K.M. et al., Mutagenic by-products from chlorination of humic acid, *Environ. Health Perspect.*, 69, 101, 1986.
74. Meier, J.R., Knohl, R.B., Coleman, W.E., Ringhand, H.P., Munch, J.W. et al., Studies on the potent bacterial mutagen, 3-chloro-4-(dichloromethyl)-5-hydroxy-2(5H)-furanone: aqueous stability, XAD recovery and analytical determination in drinking water and in chlorinated humic acid solution, *Mutat. Res.*, 189, 364, 1987.
75. Revis, N.W., McCauley, P., Bull, R., and Holdsworth, G., Relationship of drinking water disinfectants to plasma cholesterol and thyroid hormone levels in experimental studies, *Proc. Natl. Acad. Sci. U.S.A.*, 83, 1485, 1986.
76. Haseman, J.K., Crawford, D.D., Huff, J.E., Boorman, G.A., and McConnell, E.E., Results from 86 two-year carcinogenicity studies conducted by the National Toxicology Program, *J. Toxicol. Environ. Health*, 14, 621, 1984.
77. Ward, J.M., Morphology of hepatocellular neoplasms in B6C3F1 mice, *Cancer Lett.*, 9, 319, 1980.
78. Harris, R.H., *Implications of Cancer-Causing Substances in Mississippi River Water*, Environmental Defense Fund, Washington, D.C., 1974.
79. Page, T., Harris, R.H., and Epstein, S.S., Drinking water and cancer mortality in Louisiana, *Science*, 193, 55, 1976.
80. Crump, K.S. and Guess, H.A., Drinking water and cancer: review of recent epidemiological findings and assessment of risks, *Annu. Rev. Public Health*, 3, 339, 1982.
81. Craun, G.F., Epidemiologic considerations for evaluating associations between the disinfection of drinking water and cancer in humans, in *Water Chlorination: Chemistry, Environmental Impact and Health Effects*, Vol. 5, Jolley, R.L., Bull, R.J., Davis, W.P. et al., Eds., Lewis Publishers, Chelsea, MI, 1985, 133.
82. Craun, G.F., Epidemiologic studies of organic micro-pollutants in drinking water, *Sci. Total Environ.*, 47, 461, 1985.
83. Shy, C., Chemical contamination of water supplies, *Environ. Health Perspect.*, 62, 399, 1985.
84. Alvanja, M., Goldstein, I., and Susser, M., Case-control study of gastrointestinal and urinary tract cancer mortality and drinking water chlorination, in *Water Chlorination: Environmental Impact and Health Effects*, Vol. 2, Jolley, R.L., Gorchev, H., and Hamilton, D.H., Jr., Eds., Ann Arbor Science Publishers, Ann Arbor, MI, 1978, 395.
85. Gottlieb, M.S., Carr, J.K., and Clarkson, J.R., Drinking water and cancer in Louisiana: a retrospective mortality study, *Am. J. Epidemiol.*, 116, 652, 1982.
86. Struba, R.J., Cancer and Drinking Water Quality, thesis, University of North Carolina, Chapel Hill, NC, 1979.

87. Brenniman, G.R., Vasilomanolakis-Lagos, J., Amsel, J. et al., Case-control study of cancer deaths in Illinois communities served by chlorinated or nonchlorinated water, in *Water Chlorination: Environmental Impact and Health Effects*, Vol. 3, Jolley, R.L., Brungs, W.A., Cumming, R.B. et al., Eds., Ann Arbor Science Publishers, Ann Arbor, MI, 1980, 1043.
88. Young, T.B., Kanarek, M.S., and Tsiatis, A.A., Epidemiologic study of drinking water chlorination and Wisconsin female cancer mortality, *J. Natl. Cancer Inst.*, 67, 1191, 1981.
89. Cantor, K.P., Hoover, R., Hartge, P. et al., Drinking water source and bladder cancer: a case-control study, in *Water Chlorination: Chemistry, Environmental Impact and Health Effects*, Vol. 5, Jolley, R.L., Bull, R.J., Davis, W.P. et al., Eds., Lewis Publishers, Chelsea, MI, 1985, 145.
90. Cantor, K.P., Hoover, R., Hartge, P. et al., Bladder cancer, drinking water source, and tap water consumption: a case-control study, *J. Nat. Cancer Inst.*, 79, 1269, 1987.
91. Cantor, K.P., Hoover, R., Hartge, P. et al., Bladder cancer, tap water consumption, and drinking water source, in *Water Chlorination: Chemistry, Environmental Impact and Health Effects*, Vol. 6, Jolley, R.L., Condie, L.W., Johnson, J.D. et al., Eds., Lewis Publishers, Chelsea, MI, 1990, 411.
92. Cragle, D.L., Shy, C.M., Struba, R.J., and Stiff, E.J., A case-control study of colon cancer and water chlorination in North Carolina, in *Water Chlorination: Chemistry, Environmental Impact and Health Effects*, Vol. 5, Jolley, R.L., Bull, R.J., Davis, W.P. et al., Eds., Lewis Publishers, Chelsea, MI, 1988, 153.
93. Lynch, C.F., VanLier, S., and Cantor, K.P., A case-control study of multiple cancer sites and water chlorination in Iowa, in *Water Chlorination: Chemistry, Environmental Impact and Health Effects*, Vol. 6, Jolley, R.L., Condie, L.W., Johnson, J.D. et al., Eds., Lewis Publishers, Chelsea, MI, 1990, 387.
94. Wilkins, J.R. and Comstock, G.W., Source of drinking water at home and site-specific cancer incidence in Washington County, Maryland, *Am. J. Epidemiol.*, 114, 178, 1981.
95. Lawrence, C.E., Taylor, P.R., Trock, B.J., and Reilly, A.A., Trihalomethanes in drinking water and human colorectal cancer, *J. Natl. Cancer Inst.*, 72, 563, 1984.
96. Young, T.B., Wolf, D.A., and Kanarek, M.S., Case-control study of colon cancer and drinking water trihalomethanes in Wisconsin, *Int. J. Epidemiol.*, 16, 90, 1987.
97. Young, T.B., Kanarek, M.S., Wolf, D.A., and Wilson, D.A., Case-control study of colon cancer and volatile organics in Wisconsin municipal groundwater supplies, in *Water Chlorination: Chemistry, Environmental Impact and Health Effects*, Vol. 6, Jolley, R.L., Condie, L.W., Johnson, J.D. et al., Eds., Lewis Publishers, Chelsea, MI, 1990, 373.
98. Zierler, S., Danley, R.A., and Feingold, L., Type of disinfectant in drinking water and patterns of mortality in Massachusetts, *Environ. Health Perspect.*, 69, 275, 1986.
99. Zierler, S., Feingold, L., Danley, R.A., and Craun, G., Bladder cancer in Massachusetts related to chlorinated and chloraminated drinking water: a case-control study, *Arch. Environ. Health*, 43, 195, 1988.
100. Zierler, S., Feingold, L., Danley, R.A., and Craun, G.F., A case-control study of bladder cancer in Massachusetts among populations receiving chlorinated and chloraminated drinking water, in *Water Chlorination: Chemistry, Environmental Impact and Health Effects*, Vol. 6, Jolley, R.L., Condie, L.W., Johnson, J.D. et al., Eds., Lewis Publishers, Chelsea, MI, 1990, 399.

101. Murphy, P.A. and Craun, G.F., A review of previous studies reporting associations between drinking water disinfection and cancer, in *Water Chlorination: Chemistry, Environmental Impact and Health Effects*, Vol. 6, Jolley, R.L., Condie, L.W., Johnson, J.D. et al., Eds., Lewis Publishers, Chelsea, MI, 1990, 361.
102. International Agency for Research on Cancer, *Chlorinated Drinking-Water; Chlorination By-Products; Some Other Halogenated Compounds; Cobalt and Cobalt Compounds*, Vol. 52, World Health Organization, Geneva, 1991.
103. Flaten, T.P., Chlorination of drinking water and cancer incidence in Norway, *Int. J. Epidemiol.*, 21, 6, 1992.
104. McGeehin, M., Reif, J., Becker, J., and Mangione, E., A case-control study of bladder cancer and water disinfection in Colorado, Abstract 127, Society for Epidemiologic Research, Annual Meeting, Minneapolis, June 9-12, 1992.
105. Ijsselmiuden, C.B., Gaydos, C., Feighner, B., Novakoski, W.L. et al., Cancer of the pancreas and drinking water: a population-based case-control study in Washington County, Maryland, *Am. J. Epidemiol.*, 136 (7), 836, 1992.
106. Morris, R.D., Audet, A.M., Angelillo, I.F. et al., Chlorination, chlorination by-products and cancer: a meta-analysis, *Am. J. Public Health*, 82, 955, 1992.
107. Editorial, Chlorinated water and cancer: is a meta-analysis a better analysis? *Pediatr. Alert*, 17, 91, 1992.
108. Lynch, C.F., Woolson, R.F., O'Gorman, T., and Cantor, K.P., Chlorinated drinking water and bladder cancer: effect of misclassification on risk estimates, *Arch. Environ. Health*, 44, 252, 1989.
109. Zeighami, E.A., Watson, A.P., and Craun, G.F., Serum lipid levels in neighboring communities with chlorinated and nonchlorinated drinking water, in *Water Chlorination: Chemistry, Environmental Impact and Health Effects*, Vol. 6, Jolley, R.L., Condie, L.W., Johnson, J.D. et al., Eds., Lewis Publishers, Chelsea, MI, 1990, 421.
110. Zeighami, E.A., Watson, A.P., and Craun, G.F., Chlorination, water hardness, and serum cholesterol in forty-six Wisconsin communities, *Int. J. Epidemiol.*, 19, 49, 1990.
111. Riley, T.J., Cauley, J.A., Murphy, P., and Black, D., The relationship of water chlorination to serum lipids in elderly white women, Abstract 129, Society for Epidemiologic Research, Annual Meeting, Minneapolis, June 9-12, 1992.
112. Lubbers, J.R., Chausan, S., and Bianchine, J.R., Controlled clinical evaluations of chlorine dioxide, chlorite, and chlorate in man, *Environ. Health Perspect.*, 46, 57, 1982.
113. Wones, R.G. and Glueck, C.J., Effects of chlorinated drinking water on human lipid metabolism, *Environ. Health Perspect.*, 69, 255, 1987.
114. Wones, R.G., Mieczkowski, L., and Frohman, L.A., Chlorinated drinking water and human lipid and thyroid metabolism, in *Water Chlorination: Chemistry, Environmental Impact and Health Effects*, Vol. 6, Jolley, R.L., Condie, L.W., Johnson, J.D. et al., Eds., Lewis Publishers, Chelsea, MI, 1990, 301.
115. Wones, R.G., Deck, C.C., Stadler, B. et al., Lack of effect of drinking water chlorine on lipid and thyroid metabolism in healthy humans, *Arch. Environ. Health*, 1993, 99: 375-81.
116. Wones, R.G., Deck, C.C., Stadler, B. et al., Effects of drinking water monochloramine on lipid and thyroid metabolism in healthy men, *Arch. Environ. Health*, 1993, 99: 369-74.
117. Kramer, M.D., Lynch, C.F., Isacson, P., and Hanson, J.W., The association of waterborne chloroform with intrauterine growth retardation, *Epidemiologic*, 3, 407, 1992.

118. Bove, F.J., Fulcomer, M.C., Koltz, J.B., Esmart, J., Dufficy, E.M., and Zagraniski, R.T., Report on phase IV-A: public drinking water contamination and birthweight fetal deaths, and birth defects, a cross-sectional study, New Jersey Department of Health, 1992.
119. Bove, F.J., Fulcomer, M.C., Koltz, J.B., Esmart, J., Dufficy, E.M., Zagraniski, R.T., and Savrin J.E., Report on phase IV-B: public drinking water contamination and birthweight and selected birth defects, a case-control study, New Jersey Department of Health, 1992.
120. Aschengrau, A., Zierler, S., and Cohen, A., Quality of community drinking water and the occurrence of late adverse pregnancy outcomes, *Arch. Environ. Health*, 48 (2), 105, 1993.
121. Murphy, P.A., Quantifying risks from epidemiological studies: application to the disinfectant by-product issue, in *Safety of Water Disinfection: Balancing Chemical and Microbiol Risks*, Craun, G.F., Ed., ILSI Press, Washington, D.C., 1993, 373-88.

CHAPTER 9

Injury

Leon S. Robertson

Injury leads all maladies as a cause of lost potential years of life in the U.S. Although cardiovascular diseases and cancers kill larger numbers of people, most people who die from those diseases are elderly and have far fewer potential years left than the fatally injured. Among the fatally injured in 1985, the average potential life remaining was 36 years, compared to 12 years among those who died of cardiovascular diseases and 16 years among those who died of cancers. In fact, injuries claim more years of potential life annually than heart diseases and cancers combined.[1]

Besides about 142,000 injury-caused deaths a year, nonfatal injuries are an enormous burden — about 54 million physician visits and around 2.3 million hospitalizations are attributed to injury annually in the U.S. Thousands of people are permanently disabled, and 127 million days are spent in bed as a consequence of injury. The cost in dollars for lost productivity, medical care, rehabilitation, etc., not counting pain and suffering, was about $158 billion in 1985, discounted to present value in that year.[1]

One measure of the impact of particular types of injury is the rate per population. In Table 1, the deaths, hospitalizations, and estimated total numbers of injuries per 100,000 population in 1985 are presented.[1] Motor vehicle injuries are the leading cause of injury deaths, whereas falls are the leading cause of injury hospitalizations. Firearms are the second leading cause of injury deaths, but are third behind falls and motor vehicles as a cause of hospitalizations. An important factor to be considered when investigating injuries or considering injury control efforts is severity. Motor vehicles, firearms, falls, poisonings, fire/burns, and drownings account for 80% of deaths, but only 36% of nonhospitalized injuries.[2]

Table 1 is one example of various ways to describe injuries. The evolution in thought about injuries is partly reflected in how they have been classified. Injuries traditionally were classified as accidental or intentional. Injury control was primarily focused on "accident prevention" and interpersonal or self-directed violence was left to law enforcement, psychiatry, social workers, the clergy, etc. That is not to say that these professions have not contributed to injury control, or that all attempts to prevent

0-87371-573-X/95/$0.00+$.50

Table 1 Numbers and Rates of Injury Per 100,000 Population By Type and Severity of Injury, U.S., 1985

	Fatal		Hospitalized		Nonhospitalized	
	No.	Rate	No.	Rate	No.	Rate
Motor vehicles	45,923	19.4	523,028	220.6	4,803,000	2,026
Falls	12,866	5.4	783,357	330.5	11,493,000	4,848
Firearms	31,556	13.3	65,129	27.5	171,000	72
Poisonings	11,894	5.0	218,554	92.2	1,472,000	621
Fires/burns	5,671	2.4	54,379	22.9	1,403,000	592
Drownings	6,171	2.6	5,564	2.3	26,000	11
Other	28,487	12.0	696,707	293.9	35,001,000	14,765
Total	142,568	60.1	2,346,735	990.0	54,369,000	22,935

Adapted from Rice, D.P. et al., *Cost of Injury in the United States: A Report to Congress*, University of California Institute for Health and Aging, San Francisco; and The Johns Hopkins University Injury Prevention Center, Baltimore, 1989.

"accidents" have been unsuccessful, but the extent of scientific investigation of the effectiveness of these approaches was very limited.

"Accidents" refer to a very large and fuzzy set of events, only a small proportion of which are injurious. Any unintended, incidental event that interferes in one's daily pursuits is an accident. It should be clear from Table 1 that attempts to prevent all "accidents" would result in a disproportionate emphasis on the factors that contribute to the large number of rather trivial events, perhaps to the neglect of the less frequent but more often fatal or permanently disabling events. The word "accident" is also intertwined with the notion that some human error or behavior is responsible for most injuries. This focus of attention on the human actors involved tends to detract from an examination of the full range of factors that contribute to injuries and, particularly, their severity.

Accident is primarily a euphemism for lack of intent, as though intent were a primary consideration in injury prevention. If two people have an argument that results in a brief exchange of fisticuffs, the incident usually goes unrecorded as an injury. In a similar situation, if one of the persons has a gun and kills the other with it, the case is classified as homicide, as though the person intended to kill, which is often not true. It is the lethality of the weapon that determines severity, not necessarily the intent of the assailant.

The term "injury" or specific types of injury — amputation, burn, laceration, etc. — is a more precise description of the harm to be reduced. Also, when referring to attempts to reduce injury, the term "injury control" is used here. This term is preferred to injury prevention because it includes attempts at severity reduction through management of injury agents as well as medical care and rehabilitation that help reduce death and disability.

I. THE EPIDEMIOLOGICAL MODEL

The theoretical model developed by infectious disease epidemiologists is also applicable to injury. The core concepts of this model include the host (the person injured), the agent that injures, the vector or vehicle that may convey the agent, as well as other environmental factors.

What came to be known as infectious diseases were first found correlated to seasons, water sources, socioeconomic status of the populations primarily affected, and the like. We now know that in some cases these correlates of the diseases were carriers (vehicles or vectors) of infectious agents. In others they were factors that increased or decreased host exposure or susceptibility to the agents, and some were spurious.[3]

Microbiologists subsequently identified tiny biological structures (bacteria, parasites, viruses) that secreted toxins in an invaded host, or removed elements from the host, or caused other changes at the cellular or organ levels, that resulted in sickness and death. Then epidemiologists knew what to look for in the seasons, water, living conditions, etc., associated with a given disease.

In some cases the microorganism was conveyed to the host in inanimate media, such as water and milk, which came to be called vehicles. Others were carried to human hosts by insects, by animals, or were directly transmitted from human being to human being. These animate carriers came to be called vectors. Living conditions, often related to economic status, increased or decreased exposure to the carriers of the agents or increased susceptibility to infection due to nutritional or other factors.

All of these discoveries had implications for control of infectious diseases. In some cases susceptibility could be reduced by modifying the immune mechanisms of the potential hosts. In others, antibiotic agents could be introduced into the infected hosts to reduce the severity of the illness. Elimination or control of certain carriers, such as rodents and insects, could be tried and, in some cases, accomplished. Improved living conditions also reduced exposure to harmful agents or carriers. Also, epidemiologic evidence on transmission media, as well as times, places, and populations involved, was crucial in the deliberate attempts to prevent or treat many infectious diseases.

Although injuries can be characterized using the concepts of infectious disease epidemiology, injury epidemiology lagged in development by decades. The 20th century was almost two thirds past before the agents of injury were accurately identified as the various forms of energy — mechanical, thermal, chemical, electrical, ionizing radiation — or too little energy in the case of asphyxiation, and that identification came from a psychologist, not an epidemiologist. Before and since that date, certain authors referred to motor vehicles, guns, alcohol, etc. as agents of injury, but that is inaccurate in the epidemiologic use of the concept of agent. Motor vehicles and guns are vehicles of mechanical energy in epidemiologic parlance, and alcohol contributes to injury by sometimes affecting behavior that places people at greater risk of injurious energy exposure as well as perhaps increasing vulnerability of tissues to energy insults.

Prior to these insights, the research on injury was primarily focused on human characteristics and human behavior correlated with injury incidence and, more rarely, severity — with occasional studies of seasonal and geographical variations and the like. A few isolated researchers looked at human tolerance of mechanical energy as important.

Table 2 The Haddon Matrix With Examples

	Factors		
Phases	**Human**	**Vehicle**	**Environment**
Preinjury	1. Alcohol intoxication	4. Braking capacity of motor vehicles	7. Visibility of hazards
Injury	2. Resistance to energy insults	5. Sharp or pointed edges and surfaces	8. Flammable building materials
Postinjury	3. Hemorrhage	6. Rapidity of energy reduction	9. Emergency medical response

It is not that the characteristics of the energy were unknown. The leading source of injury by far is mechanical energy, the characteristics of which were known since Sir Isaac Newton's work on the laws of motion in the 17th century. Although Newton's laws of motion do not apply to the speed of light, they are applicable to motor vehicles, bullets, and falling human beings.

II. FACTORS AND PHASES OF INJURY

The transfer of energy to human beings at rates and in amounts above or below the tolerance of human tissue is the necessary and specific cause of injury. The amount of the energy concentration outside the bands of tolerance of tissue determines the severity of the injury.

Injury usually refers to the damage to cells and organs from energy exposures that have relatively sudden, discernible effects, although some researchers have included damage from chronic low-energy exposures, such as back strain or carpel tunnel syndrome, as injury. Chemical and radiation exposures that produce cellular changes resulting in neoplasms are usually called cancerous rather than injurious. A debate regarding the appropriateness of inclusion or exclusion of any harmful condition as injury or disease would be pointless, but at the fuzzy edges of a set of harmful consequences from energy exchanges, a given researcher should make clear the cases that are considered injuries.

Most of the concentrations of energy involved in severe and fatal injuries are the result of modification by human organizations of the energy inherent in matter. Some falls occur from heights unmodified by human construction. A few people are struck by lightening, tornadoes, etc. But motor vehicles and guns, as well as cigarettes, which ignite more housefires than any other ignition source, and home swimming pools, in which more children drown than in other bodies of water, are examples of human inventions that are some of the major sources of serious injuries.

To alert researchers to the factors contributing to injury incidence and severity, and the timing of involvement of those factors, Haddon[4] devised a matrix of broad categories of factors and phases of injury. This matrix, along with some examples of factors known to be important in each cell, is shown in Table 2. Prior to injury, human, vehicle, and environmental factors contribute to the increase or decrease in exposure to potentially damaging energy. During the energy exchange, the susceptibility of the host's tissue to damage and the concentration of the energy on the host by vehicle and environmental characteristics are major factors in severity. After the initial energy exchange, the condition of the host, the potential for more energy exposure, and the responses from the environment substantially affect survival and the time and extent of return to preinjury functioning of those who survive.

Haddon was also the originator of a systematic classification of technical interventions to reduce injury incidence or severity.[4] The interventions are applicable to environmental hazards generally, but possibilities that have been considered as relevant to injury control are noted here.

1. *Prevent the creation of the hazard in the first place.* Do not allow the manufacture of particularly hazardous vehicles, such as motorcycles, minibikes, and "all terrain" vehicles. These vehicles are used mainly for recreation, for which there are numerous alternative and less hazardous activities.
2. *Reduce the amount of the hazard brought into being.* Require that passenger vehicles, particularly utility vehicles, have lower centers of gravity or wider track width such that track width divided by twice the height of center of gravity (T/2H) is not less than 1.2. Vehicles with lower T/2H have 3 to 20 times as many fatal rollover crashes as those 1.2 or greater and the relative risk is strongly correlated to T/2H.
3. *Prevent the release of the hazard that already exists.* Keep guns for target shooting at the shooting range rather than in homes.
4. *Modify the rate or spatial distribution of release of the hazard from its source.* Prohibit automatic and semi-automatic guns.
5. *Separate, in time or space, the hazard and that which is to be protected.* Build pedestrian and bicycle paths separated from roads.
6. *Separate the hazard and that which is to be protected by interposition of a material barrier.* Place fences with gates that children cannot open around swimming pools and other small bodies of water in areas where children can reach them.
7. *Modify basic relevant qualities of the hazard.* Use energy-absorbing materials of adequate depth on playground surfaces.
8. *Make what is to be protected more resistant to damage from the hazard.* Develop treatment of persons with hemophilia and osteoporosis to increase resistence to mechanical energy exchanges.
9. *Begin to counter the damage already done by the environmental hazard.* Increase use of smoke detectors and carbon monoxide detectors.
10. *Stabilize, repair, and rehabilitate the object of the damage.* Provide prosthetic devices for amputees and wheelchairs, beds, and equipment used in work and other activities designed to optimize normal living.

III. INJURY DATA

The allocation of resources to injury control would be more effective if data on injuries better reflected the circumstances. Several governmental and private agencies maintain data systems that continually or periodically collect data on injuries that are used to measure trends, detect clusters, and identify factors related to injury.

These are collectively called injury surveillance systems. Some of these data systems are also good sources of cases for more detailed study.

The International Classification of Diseases[5] includes two types of codes for injury, diagnosis codes — sometimes called N-codes — and codes for "external causes", called E-codes. Even in hospitals where the injury coding is done systematically, the E- codes are often missing, particularly in cases with so many N-codes that there is no room for E-codes on computer files, as the data are currently structured. In many cases, the medical history and notes by physicians and nurses in hospital records do not contain enough detail about the circumstances of an injury to classify it by E-code or specific geographic location.

Universal E-coding of hospital discharge records has been advocated by the Council of State and Territorial Epidemiologists and a few states now require them. The Indian Health Service routinely requires such coding in its hospitals and E-codes for injuries were found on 99.3% of its injury case records, although 25% lacked sufficient information for validity of detailed three-digit codes. Comparison of the codes and detailed examination of a sample of hospital records indicated excellent reliability in general categories (motor vehicle, fall, etc.), but discrepancies increased in the more refined subcategories.

Although universal E-coding would provide much better information on trends and clusters of injuries by type, severity, and a few demographic characteristics, it would mainly serve as a source to identify cases of particular types of injury for more detailed investigation or targeting of control efforts. Cases without E-codes are identifiable as injuries by N-codes, which are much more complete.

Finding clusters by geographic areas is of major importance in targeting certain injury control efforts. Complete counts of severe injuries during a period of time, rather than samples, are required to reveal clusters, particularly if the areas are small. Usually the smaller the area, the longer the time period required for stable numbers.

IV. PREVENTION-ORIENTED SURVEILLANCE

Previous successful efforts in injury control based on surveillance have included the following steps:

1. Surveillance of injury incidence and severity to identify clusters of similar injuries and the hazards that increase incidence and severity
2. Identification of one or more technical approaches to eliminate or reduce the hazard
3. Implementation of the technical approaches among the populations at high risk
4. Continued surveillance to monitor the trend in the injuries

An outstanding example of the application of this approach occurred in the study and subsequent reduction of fatal falls to children in New York City. Epidemiologists from the health department devised a surveillance system of the circumstances of the falls and found that 66% of injuries in fatal falls to children up to 5 years of age occurred when the children crawled out of windows in high-rise buildings. The research also identified the areas of the city where these deaths most frequently occurred.

A barrier that could be placed over windows, preventing children from crawling out, was the technical approach identified as most feasible under the circumstances. A campaign was launched in high-risk neighborhoods to persuade the parents or landlords to install the barriers. Eventually, the health department required landlords to install such barriers when requested by tenants. In association with these efforts, the deaths from children's falls from high-rise windows declined from about 30 to 50 per year in the 1960s to 4 in 1980. Total reported falls declined proportionately during the same period. Subsequently, as attention to the issue declined, the falls and fatalities increased somewhat. In July 1986, the city changed the regulation to require barriers in buildings where there were children less than 11 years old.

In addition to illustration of the steps necessary for efficient injury control, the New York experience with childrens' falls from heights suggests the local nature of certain hazards. In cities and towns with few or no high-rise buildings (indeed in the boroughs of Queens and Staten Island, as the researchers found in New York), a campaign to install barriers in windows would be inappropriate because the problem is rare, relative to other types of injury. Therefore, local injury surveillance is necessary to identify major injury problems that vary widely among local areas, and their circumstances and specific locations within the areas. The local health department is an appropriate agency for such activity, but other agencies, hospitals, etc., could also do the work.

Numerous technical strategies are available for injury control, but efficient use requires data on the extent to which they are needed where the problem is most acute. For example, certain road modifications, signaling systems, and lighting reduce relevant injuries by more than 50%.[6] Yet modifying every mile of road with every possible modification would be very expensive.

By conducting detailed surveillance of the circumstances, frequency, and locations of serious injuries, the health department or other organization can recommend action to agencies or organizations in a position to implement, require, or distribute technology or other approaches. For example, if particular road intersections were found to have high rates of severe injury crashes, the data and suggestions for changes, such as extension of the yellow phase of traffic control lights at the specified intersections, would be forwarded to the road or police department that has jurisdiction. If skid strips on stairs, hand rails, lighting, or other approaches were identified as likely ameliorative strategies for specific types of falls found among the elderly, the recommendations for specific modifications could be made to vulnerable community residents by visiting nurses or other persons who provide services to the elderly.

Geographic distributions of injuries can also be used to designate the placement and staffing of emergency medical services and trauma treatment centers. For example, one emergency medical service that covered a metropolitan area of 600 square miles found that 25% of the calls occurred in two 3- by 4.5-mile areas. The severe injuries were distributed similarly.[7]

As indicated in the discussion of extant surveillance systems, few include data in sufficient detail to identify specific types of injury by specific locations, and none directly identify environmental modifications that could have reduced incidence and severity.[8] To provide such information, a supplementary data collection system was developed for the Indian Health Service.

The data to be gathered are indicated on the forms available elsewhere,[2] one form each for injury from motor vehicles, burn or smoke, drowning or near drowning, a fall, assault, suicide attempt, and others. The forms include not only the circumstances of the injury, but also a list of possible actions that might have prevented the injury or reduced severity. The form for motor vehicles is presented in Table 3 as an example. The surveillance is not oriented simply to the collection of data; it is prevention-oriented.[2]

Confining the initial effort to the more severe cases was deemed appropriate to avoid excessive effort expended on relatively trivial injuries that may occur in large numbers, but are relatively unimportant in terms of long-term consequences for the persons injured and use of community resources. The definition of "serious" is somewhat arbitrary and can be changed as progress is made in prevention of the more severe cases. Fatalities and hospitalized injuries should receive first priority.[2]

Since the Indian Health Service provides outpatient as well as inpatient and preventive services in many Native American communities, access to cases by injury prevention specialists is no doubt easier than it would be in communities with more fragmented services. Nevertheless, the potential cost savings to be obtained by targeted injury control efforts informed by data should be appealing to hospitals. Reimbursement systems based on average costs for diagnosis-related groups have resulted in insufficient payments to hospitals for certain severe injuries because of the skewed distributions of costs.

Initial experience with the use of the Indian Health Service (IHS) system indicated that lack of expertise in identifying potentially effective environmental modifications was a problem. A fellowship program to train injury control specialists and a series of seminars for other users of the system have been instituted. Technical assistance to state and local communities not served by the Indian Health Service is available from the injury prevention centers funded by the Centers for Disease Control (CDC) or from CDC. A current list of injury prevention centers is available from CDC.[9]

Computer software has been developed by the Indian Health Service that provides for easy entry of the data from the surveillance system. The program can be easily modified for use in any community. As sufficient numbers accumulate, a summary of the circumstances cross-tabulated by the suggested actions that might have had a preventive effect provides a priority list for action.[2]

Development of detailed computerized codes for injury locations to identify geographic clusters may be cumbersome, but good database management systems, such as EPIINFO used by IHS, allow listing of case identifiers by other variables. Once high-priority actions have been identified, cases that would have been reduced by a given action can be listed and the locations marked on detailed maps of local areas by referring to the location information on the original forms.[2]

Table 3 Motor Vehicle Injury Form

Community________________________________Census tract____________________
Location of the incident (specify road, street, or intersection
and distance to an identifiable reference point such as an
intersection, business, or milepost number)__________________________________

____

____

____

____

Severity: ___ fatal ___ hospitalized ___ ambulatory (fracture,
loss consciousness only — exclude others)
Age ___ Gender: M__ F__
Single vehicle occupant
__ Fixed object If fixed object: __ tree __ utility pole
__ Rollover __ bridge abutment __ light
__ Both pole __ sign pole __ other
__ Animal on the road (What? ____________________)
Multiple vehicle occupant __ Frontal __ Side __ Rear
Motorcyclist __ Single Vehicle __ Multiple vehicle
Pedestrian __ Crossing intersection __ Crossing elsewhere
__ Walking along road __ Vehicle came off road
__ Laying in road __ Other (What? ____________)
Bicyclist __ Crossing intersection __ Crossing elsewhere
__ On road parallel to traffic __ On road against traffic
__ Motor vehicle came off road
__ Other (What? ________________________)
Lighting: __ Daylight __ Dark __ Dark but lighted __ Dawn or Dusk
Signals: __ None __ Flashing Warnings __ Red-Yellow-Green
__ Stop sign __ Yield sign __ Other (What? __________________)
Crash Protection: __ Seat belt __ Child restraint __ Crash helmet
Roadway Jurisdiction: __ City or Town __ County __ State __ Fed.
Modification that might have prevented the injury or reduced
severity (check all that apply):
__ No pass stripe __ Roadside hazard removal
__ Rumble strips __ Signal or sign at intersection
__ Lengthen yellow phase at signalized intersection
__ Install or lengthen pedestrian walk signal
__ Median barrier __ Reflectors on curve
__ Snow removal __ Improve road skid resistance
__ Separate pedestrian walkway from road
__ Reflectors on vehicles or clothing
__ Lighted roadway __ Curb to limit road access
__ Other (What? ________________________________)
__ Additional observations

The system is beginning to show results. For example, injury control specialists in White River, Arizona found a cluster of 37 severe pedestrian injuries that occurred at night on a 2-mile section of road in a 2-year period. The tribal government and IHS collaborated in the installation of lights that illuminated the road section at night. Only two pedestrians were struck on that road section in the following 2 years. In Browning, Montana, 59 severe motor-vehicle injuries, including 13 fatalities, occurred in a 2-mile stretch of road during a 7-year period. Overhead lighting and curbs that channeled parking lot traffic to controlled entry points were installed. In the 2-year period after lighting and curbs were installed, only two severe injuries occurred in that stretch of road. After being shown data on a cluster of 22 fatal pedestrian injuries at night on a 2-mile section of the road between Gallup, New Mexico and the

Navajo Nation, state highway authorities put night lighting on the road section in 1992.

A visit to the site of each severe injury to consider environmental modifications that might have reduced the injuries is strongly recommended. For example, visits to the sites of child pedestrian injuries on the Pine Ridge Reservation in South Dakota indicated that the surfaces and equipment on nearby playgrounds were in such poor condition that the children apparently preferred to play in the streets or driveways of homes.[2]

The choice of recommended ameliorative actions should not necessarily be confined to the more obvious ones that can be fitted on a one-page form. The narratives and comments may suggest others. Those included on the forms are oriented to actions that can be initiated at the local level and do not include actions delegated to federal regulatory agencies. The recommended readings at the end of this chapter provide expertise in the identification of additional options for specific injuries.[1,2,4,6,8-10]

Ideally, every community would have an injury surveillance system analogous to that of IHS. If the numbers in a given community were too limited for generalization, small communities in similar areas could pool the data to assess patterns for their environment. A system for accumulating data from the local systems at the state (or provincial) and national levels would give each level of government, or private entity, information on injury patterns relevant to agencies or organizations under its purview. Since national systems may be long in coming, local communities that are concerned about their injury problems can take the initiative.

Use of the IHS or similar forms could be required of medical examiners, coroners, and hospitals. The mechanism of enforcement of quality of data from medical examiners and coroners is not evident, but hospitals could be required to obtain the data to qualify for reimbursement by Medicare, Medicaid, or private insurance.

If and when a national system is developed, the information gathered in local surveillance systems must be made uniform on certain variables. For use by national regulatory agencies and independent researchers, the specific identification of product brand names and other identifiers such as serial numbers should be included. Where structures or other facilities that are, or could be, subject to local codes and ordinances are involved, the builders or maintainers should be identified. The mere fact that the data are being collected could serve as motivation for some organizations to undertake injury control actions. The data would give them better information on actions to take.

V. CONCLUSION

Injury control is more readily accomplished when an epidemiologic perspective is employed. Focus on individual behavior, negligence, or intent ignores practical changes in energy or its carriers that can greatly reduce injury severity, if not incidence. For example, if the New York Health Department had focused on all sorts of behavior — alcohol use, no adult present, numerous conditions that distract adults from continuously watching children, hyperactive children, etc. — it is doubtful that children's falls from windows would have been reduced much, if any. Similarly, the often-heard assumption that the high injury rates among Native Americans can only be reduced by reducing alcohol use was proved false. Although some, but certainly not all, of the pedestrians injured in White River were drunk, they were not hit by motor vehicles when they could be seen after lighting was installed.

Injuries are seldom randomly distributed. They are substantially concentrated in space, time, or in certain populations. Collection of data on who, when, where, and how people are injured can lead to enormous reductions when simple and practical means are used to target particular clusters in populations, places, or at given times.

REFERENCES

1. Rice, D.P., MacKenzie, E.J. et al., *Cost of Injury in the United States: A Report to Congress*, University of California Institute for Health and Aging, San Francisco; and The Johns Hopkins University Injury Prevention Center, Baltimore, 1989.
2. Robertson, L.S., *Injury Epidemiology*, Oxford University Press, New York, 1992.
3. Buck, C., Llopis, A., Najera, E., and Terris, M., Eds., *The Challenge of Epidemiology: Issues and Selected Readings*, Pan American Health Organization, Washington, D.C., 1988.
4. Haddon, W., Jr., On the escape of tigers: an ecologic note, *Technol. Rev.*, 72, 44, 1970.
5. Health Care Financing Administration, *International Classification of Diseases,* 9th Revision, 3 volumes, Department of Health and Human Services, Washington, D.C., 1980.
6. Federal Highway Administration, *Synthesis of Safety Research Related to Traffic Control and Roadway Elements*, 2 volumes, U.S. Department of Transportation, Washington, D.C., 1982.
7. Pepe, P.E., Mattox, K.L., Fischer, R.P., and Matsumoto, C.M., Geographic pattern of urban trauma according to mechanism and severity of injury, *J. Trauma*, 30, 1125, 1990.
8. Baker, S.P., O'Neill, B., and Ginsberg, M., *The Injury Fact Book*, 2nd ed., Oxford University Press, New York, 1992.
9. National Committee for Injury Prevention and Control, Appendix A. Injury prevention: meeting the challenge, *American Journal of Preventive Medicine (Supplement)*, Oxford University Press, New York, 1989.
10. Robertson, L.S., *Injuries: Causes, Control Strategies and Public Policy*, D.C. Heath, Lexington, MA, 1983.

CHAPTER 10

Ionizing Radiation

Daniel J. Strom

I. INTRODUCTION

This chapter deals with environmental epidemiologic studies of the effects of ionizing radiation. For clarity, some nonionizing radiation terms are discussed for contrast with ionizing radiation.

Ionizing radiation includes electromagnetic radiation with enough energy to break chemical bonds through nonthermal processes (X and gamma photons or "rays"); and high-speed subatomic particles (electrons or beta particles, helium nuclei or alpha particles, heavy nuclei, protons, and neutrons). Ionizing radiation can be produced by electrical devices such as X-ray machines. Ionizing radiation also can arise from nuclear processes such as radioactivity (alpha, beta, and gamma emissions commonly are associated with radioactivity), nuclear fission and fusion, and from outer space.

Nonionizing radiation is electromagnetic radiation with insufficient energy to break chemical bonds through nonthermal processes. Examples of nonionizing radiation include radio and TV transmissions; microwaves and radar; and infrared, visible, and ultraviolet light (including that produced by lasers). Nonionizing radiation is also known as electromagnetic fields (EMFs). A representation of the *electromagnetic spectrum* is shown in Figure 1.

By necessity, this chapter contains a lot of background material on radiation, including some radiation terms and concepts and a catalogue of the known effects of radiation, before discussing study design concepts and recent research.

II. BACKGROUND

In the ancient art and recent science of toxicology, ionizing and nonionizing radiations are relative newcomers. The ancients knew of lodestones (natural mag-

0-87371-573-X/95/$0.00+$.50

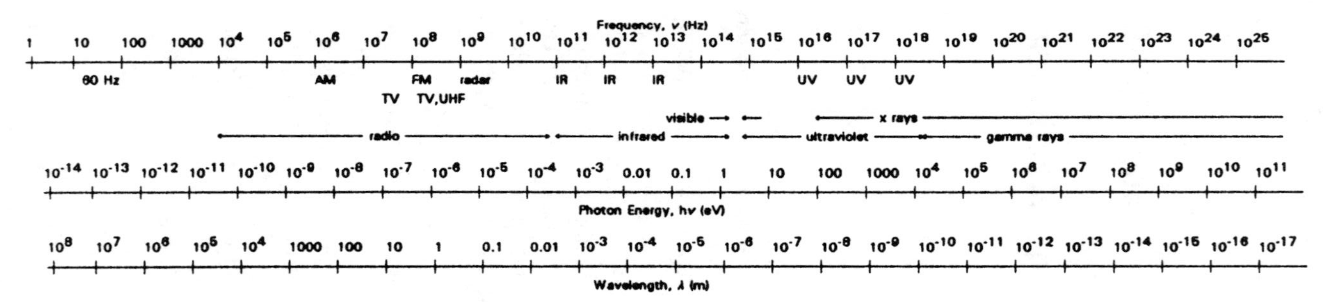

Figure 1 The electromagnetic spectrum.

nets), static electricity (the Greek word *elektron* means "amber", a substance that can easily be charged electrostatically), and lightning; the Sumerians even had crude electric cells. The ancients knew that lightning, the thunderbolt of the gods, could kill or injure. But it was not until the 19th century that scientists began to understand electric and magnetic *fields* and their extent in space, and the propagation of "Hertzian waves" (electromagnetic radiation, including radio waves) was elucidated by Maxwell in 1873.

X-rays were discovered in 1895 by Röntgen, radioactivity and its associated radiations in 1896 by Becquerel. Soon afterward, the harmful effects of ionizing radiation became known, and controversies about its risks and benefits began.

III. RADIATION TERMS

A review of ionizing radiation and radioactivity terms has recently been published by Kathren and Peterson.[1]

A. RADIOACTIVITY TERM

Radioactivity is the phenomenon of spontaneous changes (quantum-mechanical transitions) in atomic nuclei, usually accompanied by the emission of ionizing radiation and transmutation of an atom of one element into an atom of another. Since such transitions happen on an atom-by-atom basis, the unit of *amount of radioactive material* is a *rate* at which atoms are making transitions. The unit of radioactivity is the becquerel (1 Bq = 1 nuclear transition per second). An old unit is the curie (1 Ci = 37,000,000,000 Bq).

The *concentration of radioactive material* in air is usually measured in units of activity per unit volume (Bq m^{-3}). Concentrations in food, water, or soil are usually given in activity per unit mass (Bq kg^{-1}). In the U.S., concentrations of indoor radon are often quoted in units of picocuries per liter (1 pCi L^{-1} = 37 Bq m^{-3}). A traditional unit for concentration of the short-lived decay products of radon is the working level (1 WL results from 7400 Bq m^{-3} of radon in 50% equilibrium, a common indoor value). A "dose"-like unit is the working level month (WLM), equivalent to breathing 1 WL for 1 working month of 170 h.

B. IONIZING RADIATION DOSE TERMS

An *absorbed dose rate* can be measured for ionizing radiation field emitted from an X-ray machine or a radioactive source. The unit of absorbed dose rate is the gray per hour (Gy h^{-1}). One gray* is the depositing of one joule of ionizing radiation energy per kilogram of material (1 Gy = 1 J kg^{-1}). Common submultiples of the gray are the milligray (1 mGy = 10^{-3}) and microgray (1 μGy = 10^{-6} Gy). A person standing

* A traditional unit of absorbed dose is the rad. 1 rad = 0.01 Gy. Since the rad is not an SI unit, most scientific journals and international publications no longer use it, although it is still in use in the U.S.

in a radiation field of 2 mGy h^{-1} for 3 h would receive an *absorbed dose* of $2 \times 3 =$ 6 mGy.

A quantity related to the absorbed dose is *dose equivalent*, which is a quantity used in ionizing radiation protection. The unit of dose equivalent is the sievert* (Sv). Dose equivalent in sieverts is related to absorbed dose in grays by the *quality factor*, Q. The quality factor is generally taken as 1 for X-ray and gamma photons and β particles, 20 for α particles, 3 for thermal neutrons, and 20 for fast neutrons. Thus, for α particles and fast neutrons, 1 mGy produces a dose equivalent of 20 mSv; for thermal neutrons, 1 mGy produces 3 mSv; and for β, γ, and X radiation, 1 mGy produces 1 mSv. Quality factors are only valid for low doses and low dose rates, so the dose equivalent has little meaning above 1 Sv.

Quality factors are a surrogate for *relative biological effectiveness* (RBE), a quantity used to compare radiations of different types by comparing absorbed doses of each type that produce the same biological effect. Since RBE (and thus Q) should be the result of an experiment, rather than used as input in the dose variable, it would be preferable to keep different types of radiations separate and use absorbed dose of each type as the independent variables in epidemiologic studies. Indeed, the determination of the RBE (and thus Q) for neutrons has been one of the goals of the studies of the Japanese A-bomb survivors.[2] However, epidemiologic studies often use dose equivalent rather than absorbed dose as the dose variable. This is unavoidable in many circumstances, but is undesirable because little can be learned about the quality factor unless dose equivalents from different kinds of radiation are kept separate.

C. NONIONIZING RADIATION TERMS

Nonionizing radiation quantities for ambient measurements include electric field strength (volts per meter, V m^{-1}), magnetic field strength (amperes per meter, A m^{-1}), magnetic flux density (webers per square meter [Wb m^{-2}] or tesla** [T]), and polarization of these fields. Frequency in hertz (Hz), of course, is key to characterizing exposures to nonionizing radiation, and harmonic analysis of the distribution of energy as a function of frequency may be necessary. The specific absorption rate (SAR) is the power absorbed per unit mass of tissue (W kg^{-1}), and is known to be related to heating, but often does not correlate well with biological effects. The time integral of SAR is the specific absorbed energy (J kg^{-1}), analogous to absorbed dose for ionizing radiation, but possibly irrelevant. As discussed below, the relevant "dose" quantity for nonthermal effects of nonionizing radiation is not clearly known.

IV. KNOWN HEALTH EFFECTS OF IONIZING RADIATION

First to be discovered were the acute or prompt effects of ionizing radiation.

* A traditional unit of dose equivalent is the rem. 1 rem = 0.01 Sv. Since the rem is not an SI unit, most scientific journals and international publications no longer use it, although it is still in use in the U.S.

** A traditional unit of magnetic flux density is the gauss. 1 gauss = 0.0001 T. Since the gauss is not an SI unit, most scientific journals and international publications no longer use it, although it is still in use in the U.S. The earth's magnetic field is about 0.5 milligauss or 50 μT.

Table 1 Doses, Effects, and Levels of Exposure to Ionizing Radiation

Dose[a]	Reference Level *(Italics)* or Effect of Acute Whole Body Exposure to this Dose
4 190 000	1°C temperature rise (direct heating)
1 000 000	*Bottom end of food irradiation scale*
300 000	Cerebrovascular (CV) syndrome
100 000	"Prompt, immediate incapacitation" — U.S. Army, neutron bomb
30 000	
10 000	Gastrointestinal (GI) syndrome
3 000	$LD_{50/30}$ 50% die 30 d no medical care; bone marrow syndrome
1 000	Mild clinical symptoms in some; increased risk of cancer, genetic effects
300	No clinical symptoms; chromosome aberrations
100	
30	*50 mSv/year = current occupational limit*
10	*Partial body dose from lumbar spine procedure; nuclear medicine*
3	*Dental X-ray; avg. annual rad. worker (2.3); avg. annual natural (3 mSv)*
1	*Average U.S. medical + dental dose each year = 0.56 mSv*
0.3	
0.1	Partial body dose from chest X-ray

Note: Vertical scale is roughly logarithmic, covering almost 8 orders of magnitude.

[a] For values above 1000, units are mGy; below 1000, units are mSv.

Acute effects appear within hours of irradiation, and their severity is a function of the amount of radiation exposure received, the absorbed dose. A listing of radiation doses and effects is given in Table 1. Effects are discussed in detail by the National Academy of Sciences[2,3] and the United Nations Scientific Committee on the Effects of Atomic Radiation.[4] It is useful to divide the biological health effects of radiation into two categories: *somatic effects* and *genetic effects*.

Somatic effects are effects occurring in the exposed person and, in turn, may be divided into two classes: *prompt effects* that are observable soon after a large or acute dose (for example, 1 Sv or more to the whole body within a few hours); and *delayed effects*, such as cancer, that may be observable only years after exposure to radiation, or radiation dermatitis that may be caused by moderately large, chronic exposures to radiation. *Teratogenic effects* are a special case of somatic effects: they are effects that may be observed in children who were exposed during the fetal and embryonic stages of development (that is, effects to unborn children).

High-dose radiation effects, known as deterministic effects, are listed in Table 2. Among these is a collection of symptoms known as the "acute radiation syndrome", which includes skin reddening, hair loss, nausea, and vomiting, etc., and has a threshold of more than 1 Gy for acute exposures and several times higher for chronic exposures. High dose effects are only of concern after radiation accidents or detonation of nuclear weapons. There were no high-dose effects around Chernobyl except among those responding to the emergency, primarily fire-fighters.

At lower doses, we are concerned primarily with cancer and genetic effects. Genetic effects (sometimes called "heritable effects") are abnormalities that may occur in the future children of exposed individuals and in subsequent generations. They are stochastic effects, whose frequency in a large population is a function of dose, often without threshold.

Low doses of radiation are also believed to cause cancer. Cancer dose-response

Table 2 High-Dose, Deterministic Radiation Effects

- Deterministic effects exhibit a dose threshold: not seen below 0.1 Sv/0.1 Gy
- The *severity* of the effect depends on *magnitude of dose*
- DNA is the principal target for radiation-induced cell killing
- Biological consequence for a given absorbed dose varies with radiation "quality", which is measured as linear energy transfer (LET)
- Dose-response relationship for cell killing depends on LET

LOW LET (X, beta, and gamma):	$k_1 \times \text{dose} + k_2 \times \text{dose}^2$	(linear-quadratic)
HIGH LET (alpha, neutron)	$k_1 \times \text{dose}$	(linear)

Examples of deterministic radiation effects (some specific to radiation, others not):

Lethal (prompt)
- Necrosis or cell killing (local irradiation)
- Bone marrow (hematopoietic) syndrome: may be fatal
- GI (gastrointestinal) syndrome: fatal without extreme medical care
- CV (cerebrovascular) syndrome: invariably fatal
- Skin burns (beta burns), blisters, necrosis: may be fatal

Prompt and delayed sublethal deterministic effects
- Erythema (skin reddening)
- Nausea, vomiting
- Epilation (hair loss)
- Leukopenia (lowered white blood cell count)
- Fatigue
- Sterility (temporary or permanent)
- Cataract of the lens of the eye
- Radiation dermatitis
- Developmental abnormalities in the embryo/fetus (threshold 0.2 Gy): microcephaly, mental retardation, deformities, growth retardation, miscarriages

relationships are multiplicative (or relative) risk models, that is, an exposure to radiation multiplies the pre-existing cancer risk. Cancers seen in excess in the Japanese A-bomb survivors and other human populations exposed to relatively high doses of ionizing radiation include leukemia, especially myeloid (but never chronic lymphocytic), and solid tumors: bone surfaces, thyroid, liver, lung, bladder, breast, stomach, colon, esophagus, and skin.[5] Smoking is known to interact with radiation in a multiplicative or at least superadditive way.[2,3]

Thus, besides causing acute radiation syndrome, radiation is a known mutagen, carcinogen, and teratogen.

Epidemiologic researchers must be aware that in addition to causing biological changes, and perhaps damage or harm, radiation is associated with psychological, psychosomatic, and behavioral effects. Few agents are as feared as radiation.[2,6] Public fear of radiation and its associated psychological effects are the only documented health effect of the Three Mile Island nuclear accident.[2,5,7,8] The one unequivocal health finding following the Chernobyl nuclear accident of 1986 is "substantial negative psychological consequences in terms of anxiety and stress due to the continuing and high levels of uncertainty, the occurrence of which extended beyond the areas of concern".[9,28] These psychological problems were "wholly disproportionate to the biological significance of the radioactive contamination" and "are prevalent even in the surveyed control settlements [that received minimal exposure]".[9] It is widely believed that there was a dramatic rise in elective abortions in the vicinity around Chernobyl;[10] if this can be substantiated, it is a major health impact. Collins has investigated effects on the mental health of persons at the Kyshtym, Chelyabinsk, and Chernobyl nuclear accidents.[26] Collins and de Carvalho have documented stress in persons involved in the 1987 Goiânia, Brazil, ^{137}Cs accident.[27] Koscheyev and co-workers have documented behavior changes in Chernobyl workers from radiation-induced stress using the MMPI test.[29] Quadrel and co-workers have used the mental models approach to investigate worker attitudes towards radiation and have suggested using this approach for environmental studies.[30]

There have been so many lawsuits involving radiation that it has been called a "litigen".[11] Psychological, psychosomatic, and behavioral effects are strongly associated with risk perception, which in turn is governed largely by access to information and heavily influenced by the media.[12]

V. METHODOLOGIC ISSUES IN ENVIRONMENTAL RADIATION EPIDEMIOLOGY

Beebe[13] has outlined methodologic issues, and this outline is expanded by the BEIR Committees[2,3] and UNSCEAR.[4]

Study design — Ecological studies of low-dose radiation effects are of little scientific value,[2,13] although they capture the popular imagination and may be politically attractive. Case-control designs and cohort designs are most suitable for studies of environmental radiation effects.

Population size — The endpoints to be examined are typically rare, so that immense numbers of subjects are needed for cohort design studies. There are well-established upper bounds on radiation risks at low doses, since so many studies have been carried out. It is clear that the effects will probably not be large compared with natural incidence, so studies have to be designed with this in mind.

Low-dose radiation health endpoints are not unique to radiation — High-dose ionizing radiation effects, such as cerebrovascular syndrome (failure of the central nervous system and blood vessel walls), GI syndrome (the temporary or permanent destruction of the lining of the gastrointestinal tract followed by infection), and hematopoietic syndrome (depression or destruction of the stem cells in the bone marrow, with subsequent temporary or permanent impairment or failure of the immune system), are unambiguously recognizable from clinical symptoms as being caused by ionizing radiation. The low-dose effects, such as cancer and genetic effects, are not unique to radiation. There is always a pre-existing incidence of cancer and genetic effects in a population before irradiation. Putative nonionizing radiation endpoints, such as leukemia and breast cancer in males, also exist in groups without exceptional exposures to EMFs; the increased relative risks of these endpoints may be elevated.

Effect modifiers — Tobacco use, especially cigarette smoking, is associated with most of the cancer endpoints of interest in environmental radiation studies. In studies where lung cancer is an endpoint, or when excess relative risks are small, there is a danger of spurious associations if tobacco use is not controlled for. For example, in uranium miner studies, virtually all of the excess lung cancers occurred in smokers.[3]

There is no "unexposed" group — Natural background radiation from terrestrial radioactive materials and cosmic radiation exposes everyone. Technologically enhanced radiation exposures, such as those due to indoor radon, tailings from uranium, phosphate, and other mines, also contribute exposures. Finally, purely human-made radiation sources, such as medical X-ray machines, nuclear power reactors, radionuclides used in medicine, and nuclear weapons, contribute exposures. The task of the environmental radiation epidemiologist is to distinguish between higher and lower exposures, rather than to distinguish between exposed and unexposed. Thus, dose assessment is vitally important for high-quality radiation epidemiology. This will help to minimize information bias.[14,15]

Suitable control populations — Socioeconomic status (SES) is an important predictor of risk for the kinds of effects associated with radiation, so the control population must be carefully matched on SES.

Dose assessment — Dose rate and dose fractionation are known to be important in biological effects, due to repair. Low doses or low dose rates are *less* effective in producing cancer than high doses delivered at high dose rates. Dose rate effectiveness factors of 2 to 10 are observed for X, beta, and gamma radiations in many experiments. Dose fractionation (splitting the dose into small increments separated in time) permits repair of radiation damage for low-LET (X, β, and γ) radiation.

Tissue dose — Nonuniform or partial-body exposures (e.g., X-rays, internal radionuclides) must be accounted for. Furthermore, attenuation by the body may be significant in cases of weakly penetrating radiation. For example, very little radiation from dental X-rays to a pregnant woman can reach her unborn child, due to the nature and direction of the radiation beam.

Bias problems in environmental epidemiology — Researchers must realize that the kinds of effects caused by radiation, that is, cancer, birth defects, and mutations, are highly dreaded by the public[12] and may introduce special problems of information bias in the form of self-reporting.

Latency and wasted radiation — Clinical symptoms of cancer take 2 years to decades to appear following exposure. This time is called the latent period. Analysis must use exposure "lagging" to compensate: doses occurring within 5 to 10 years of diagnosis are generally removed from analysis (except for leukemia, where latent periods as short as 2 years were observed in bomb survivors).

VI. CASE STUDY: INDOOR RADON

Recently, a well-designed case-control study was published of 433 female lung cancer cases and 402 female controls in New Jersey for whom smoking, race, diet, and individual radon measurements were available.[16] This study shows a statistically significant dose-response gradient. While it would be desirable to have a larger population to increase significance, this study had odds ratios of 1.1, 1.3, and 4.2 for the 1.0–1.9, 2.0–3.9, and 4.0–11.3 pCi l^{-1} groups, respectively, when compared to the 0–0.9 pCi l^{-1} group. This dose-response gradient is consistent with expectations from uranium miner studies.

In a study of 308 female lung cancer cases in China, there was no apparent association between radon concentrations above 4 pCi l^{-1} and lung cancers.[17] However, only about 20% of the women lived in houses with concentrations above 4 pCi l^{-1}. At least 14 other studies are in progress in this important area.[18]

VII. CASE STUDY: HIGH NATURAL BACKGROUND RADIATION

Studies in areas with significantly elevated natural background radiation, such as Guarapari, Brazil and Kerala, India, have hinted at effects of low doses of ionizing radiation.[2] A study of 70,000 persons in Yangjiang County, Guangdong Province, People's Republic of China, showed no associations that were not spurious.[19]

VIII. CASE STUDY: FALLOUT FROM ATMOSPHERIC NUCLEAR WEAPONS TESTS

Archer has studied leukemia rates on a regional basis as a function of time and fallout levels.[20] The results are consistent with risk rates among the Japanese A-bomb survivors.

IX. THE FUTURE

There is still much work to be done in environmental radiation epidemiology, particularly in the area of indoor radon. Clearly, an important field of study in the future will be the epidemiology of fear and psychological effects caused by environmental agents such as ionizing and nonionizing radiation, and the behavior changes that result from psychological effects.

Other probable future developments include multiple endpoints and shifts in exposure-response paradigms.

A. MULTIPLE ENDPOINTS PER PERSON

One solution to the population size problem is to use multiple endpoints for each exposed individual. The use of biomarkers, that is, changes in molecular, subcellular, or cellular structures or numbers, holds great promise. Presumably, many biomarkers can be identified that will allow quantification of response for many endpoints, perhaps even with discrimination between chemical agents and radiation, and with some differential sensitivity to repair over time.

Biomarkers may also indicate a genetic predisposition to high or low sensitivity to radiation, so that sensitive subpopulations can be identified. The recent discovery of a link between breast cancer and women who had one copy of a particular gene is a good example.

B. PARADIGM SHIFTS

In the future, we can also expect the examination of shifting paradigms of disease and exposure. Raabe and co-workers have shown a striking relationship, not between dose and response, but between average dose rate and time to death by various causes.[21] Raabe's work demonstrates an effective low-dose rate threshold below which there seems to be no excess of bone cancers in mice, dogs, and humans. If this model proves to be valid, then low-dose epidemiology may not be as important as now believed.

Another challenge to the dose-response paradigm comes from the microdosimetry researchers who believe that absorbed dose, as currently defined, has very little meaning at levels of say, 10 mGy.[22-24] If they are correct (and their arguments are persuasive), then we are not even making the correct measurements when we measure dose and dose rate at low levels. Only more research will reveal the answers.

Alternative models to those currently used in low-level radiation epidemiology are available.[21,25] Cohen has suggested three alternatives to the additive risk model and the multiplicative risk model currently used by the National Academy of Sciences and others: they are the variable latent period additive risk model, the constant rate of initiation model, and the constant rate of initiation-promotor contributing model.[25] If such models prove valid, they could change the predictions of low-level radiation epidemiology.

REFERENCES

1. Kathren, R. L. and Peterson, G. R., Units and terminology of radiation measurement: a primer for the epidemiologist, *Am. J. Epidemiol.*, 130, 1076, 1989.
2. National Academy of Sciences, Committee on the Biological Effects of Ionizing Radiations (BEIR), *Health Effects of Exposure to Low Levels of Ionizing Radiation: BEIR V*, National Academy Press, Washington, D.C., 1990.
3. National Academy of Sciences, Committee on the Biological Effects of Ionizing Radiations (BEIR), *Health Risks of Radon and Other Internally Deposited Alpha-Emitters*, National Academy Press, Washington, D.C., 1990.
4. United Nations Scientific Committee on the Effects of Atomic Radiation (UNSCEAR), *Sources, Effects, and Risks of Ionizing Radiation: UNSCEAR 1988 Report*, United Nations, available from Unipub, Albany, NY, 1988.
5. International Commission on Radiological Protection, *The 1990 Recommendations of the International Commission on Radiological Protection*, Annals of the ICRP 21(1-3), ICRP Pub. 60, Pergamon Press, Oxford, 1991.
6. Weart, S., *Nuclear Fear: A History of Images*, Harvard University Press, Cambridge, MA, 1988.

7. Collins, D. L., Baum, A., and Singer, J. E., Coping with chronic stress at Three Mile Island: psychological and biochemical evidence, *Health Psychol.*, 2(2), 149, 1983.
8. Collins, D. L., Stress at Three Mile Island: altered perceptions, behaviors, and neuroendocrine measures, in *The Medical Basis for Radiation-Accident Preparedness.* Vol. 3, *The Psychological Perspective*, Ricks, R. C., Berger, M. E., and O'Hara, F. M., Jr., Eds., Elsevier, New York, 1991.
9. International Atomic Energy Agency, *The International Chernobyl Project, An Overview: Assessment of Radiological Consequences and Evaluation of Protective Measures*. International Atomic Energy Agency, Vienna, 1991.
10. Wald, N., personal communication, 1988.
11. Gallo, M., quoted in *Science*, 251, 624, 1991.
12. Slovic, P., Layman, M., Kraus, N., Flynn, J., Chalmers, J., and Gesell, G., Perceived risk, stigma, and potential economic impacts of a high-level nuclear waste repository in Nevada, *Risk Anal.*, 11(4), 683, 1991.
13. Beebe, G. W., A methodologic assessment of radiation epidemiology issues, *Health Phys.*, 46, 745, 1984.
14. Gilbert, E. S., *The Effects of Errors in the Measurement of Continuous Exposure Variables on the Assessment of Risks*, PNL-6578, Pacific Northwest Laboratory, Richland, WA, 1988.
15. Armstron, B., The effects of measurement errors on relative risk regressions, *Am. J. Epidemiol.*, 132, 1176, 1990.
16. Schoenberg, J. B., Klotz, J. B., Wilcox, J. B., Nicholls, G. P., Gil-del-Real, M. T., Stemhagen, A., and Mason, T. J., Case-control study of residential radon and lung cancer among New Jersey women, *Cancer Res.*, 50, 6520, 1990.
17. Blot, W. J., Xu, Z.-Y., Boice, J. D., Jr., Zhao, D.-Z., Stone, B. J., Sun, J., Jing, L. B., and Fraumeni, J. G., Jr., Indoor radon and lung cancer in China, *J. Natl. Cancer Inst.*, 82(12), 1025, 1990.
18. Samet, J. M., Stolwijk, J., and Rose, S. L., Summary: International Workshop on Residential Radon Epidemiology, *Health Phys.*, 60(2), 223, 1991.
19. Tao, Z. and Wei, L., An epidemiological investigation of mutational diseases in the high background radiation area of Yangjiang, China, *J. Radiat. Res.*, 27, 141, 1986.
20. Archer, V. E., Assocation of nuclear fallout with leukemia in the United States, *Arch. Environ. Health*, 42, 263, 1987.
21. Raabe, O. G., Rosenblatt, L. S., and Schlenker, R. A., Interspecies scaling of risk for radiation-induced bone cancer, *Int. J. Radiat. Biol.*, 57(5), 1047, 1990.
22. Bond, V. P., Varma, M. N., Sondhaus, C. A., and Feinendegen, L. E., An alternative to absorbed dose, quality, and RBE at low exposures, *Radiat. Res.*, 104, S-52, 1985.
23. Goodhead, D. T., Spatial and temporal distribution of energy, *Health Phys.*, 55(2), 231, 1988.
24. Morstin, K., Bond, V. P., and Baum, J. W., Probabilistic approach to obtain hit-size effectiveness functions which relate microdosimetry and radiobiology, *Radiat. Res.*, 120, 383, 1989.
25. Cohen, B. L., Alternatives to the BEIR relative risk model for explaining atomic-bomb survivor cancer mortality, *Health Phys.*, 52, 55, 1987.
26. Collins, D. L., Behavioral differences of irradiated persons associated with the Kyshtym, Chelyabinsk, and Chernobyl nuclear accidents, *Military Med.*, 157, 548, 1992.
27. Collins, D. L., de Carvalho, A. B., Chronic stress from the Goiânia ^{137}Cs radiation accident, *Behavioral Med.*, 18, 149, 1993.

28. Ginzburg, H. M., The Psychological Consequences of the Chernobyl Accident, *Public Health Rep.*, 108, 184, 1993.
29. Koscheyev, V. S., Martens, V., Kosenkov, A., Lartsev, M., Leon, G. R., Psychological functioning of Chernobyl Nuclear Power Plant operators after the nuclear disaster, *J. Traumatic Stress*, 1993.
30. Quadrel, M. J., Blanchard, K. A., Lundgren, R. E., McMakin, A. H., Mosley, M. T., Strom, D. J., U.S. Department of Energy workers' mental models of radiation and chemical hazards in the workplace, PNL-9467, Pacific Northwest Laboratory, Richland, WA, 1994.

CHAPTER 11

Electromagnetic Fields and Cancer Risks

Gilles Thériault

It has been 13 years now since the possibility of an association between cancer and exposure to electromagnetic fields (EMF) was first raised. During those years, over 100 studies have been conducted. Although some knowledge on the interaction of living organisms with electromagnetic fields has been gained, the convincing evidence that EMF causes cancer in man is still evasive.

I. RESIDENTIAL STUDIES

Cancer was first associated epidemiologically with exposure to EMF in 1979, when Wertheimer and Leeper[1] reported that children dying from cancer resided more often in homes with high current configuration than did healthy control children. They studied 344 cancer children and compared their exposure to electrical wires and electrical equipment with 344 controls. The odds ratio that they observed was 2.23 (CI 1.56–3.18) for all cancers, 2.98 (CI 1.72–5.15) for leukemias, and 2.4 (CI 1.08–5.36) for cancers of the nervous system.

They summarized their paper as follows:

> An excess of electrical wiring configurations suggestive of high current flow was noted in Colorado in 1976-1977 near the homes of children who developed cancer, as compared to the homes of control children. The finding was strongest for children who had spent their entire lives at the same address, and it appeared to be dose-related. It did not seem to be an artifact of neighbourhood, street congestion, social class, or family structure. The reason for the correlation is uncertain; possible effects of current in the water pipes or of AC magnetic fields are suggested.[1]

These results were received with much skepticism by the scientific community, which did not believe that such an association was biologically plausible and which blamed the authors for having no direct measurement of EMF exposure. Soon, other

0-87371-573-X/95/$0.00+$.50

epidemiologists duplicated Wertheimer's study in an attempt to either confirm or reject their findings.

From 1985 to 1988, four such studies were published.[2-5] They have yielded conflicting results. Some have confirmed Wertheimer's observations, some have rejected them. The study by Savitz in 1987[4] is particularly revealing. This study had been carefully planned and was hopefully free from the methodological weaknesses blamed on the previous studies. Savitz used direct measurement of EMF inside residences as well as Wertheimer's wiring configuration code. (This code allocates houses into categories of potential current flow based on the configuration of electrical wires and equipment around the house, such as nearness and size of wires, closeness to origin of current, etc.) In his results, he did not observe an association between cancer and residential exposure when direct measurements were used; but he confirmed that childhood cancer and leukemia were associated with EMF when intensity of exposure was estimated by the wiring code. This observation was most intriguing and fostered intense debates in the scientific community. There was a need for more epidemiological research to shed some light on the issue.

In Los Angeles County, California, London et al.[6] attempted to address the same hypothesis using multiple methods of assessing exposure: measurement of magnetic fields in subjects' homes over 24 h or more, spot measurements of both electric and magnetic fields, and indirect assessment of exposure by wire configuration and self-report of appliance use. The 232 cases interviewed were newly diagnosed cases of leukemia occurring between 1980 and 1987 in children from birth to age 10 years who were participants in the Los Angeles County Cancer Surveillance Program. The matched controls were drawn from friends and by random digit dialing. This study showed no apparent association between leukemia risk and measured magnetic or electric fields. However, there was an observed association (odds ratio, OR 2.15 [C.I. = 1.08–4.28] p for trend of 0.008) between the wire configuration, as used by the Denver Wertheimer-Leeper study, and childhood leukemia; even after adjustment for potential confounders.

After 13 years, we are almost back to square one. It is now taken as granted that childhood leukemia is associated with the configuration of electrical wires around residential houses. It has not been possible to show an association with measured EMF fields inside houses. What does wire code really mean? Why are not measured fields yielding better results than wire codes?

For the moment one can only propose some hypothetical reasons. If there is a real association between cancer and EMF exposure, then wire code represents either a better measure of long-term exposure to EMF than instant or 24-h in-house measurements do (which is a reasonable assumption) or wire code is a measure of field parameters (such as high-frequency transients, spikes, resonance, combination of earth magnetic-man made magnetic fields, or the angle at which magnetic field interacts with the cell) that have not been measured so far by the use of current methods of average electric and magnetic exposure estimates. Or perhaps there is no such association and therefore, wire code could be a surrogate for confounders such as low socioeconomic status, high traffic density, more densely populated areas, etc. These hypotheses need to be confirmed. So far, studies controlling for the latter confounders have not eliminated the higher risks observed.

II. EVIDENCE FROM OCCUPATIONAL STUDIES

Since data on occupational groups are reasonably accessible, it became natural to look at electrical workers, particularly electricians and electric utility workers, to test the hypothesis of an association between cancer and exposure to EMF. There have been scores of occupational studies that can be grouped into two categories: hypothesis generating and hypothesis testing studies.

A. HYPOTHESIS GENERATING STUDIES

The author who first reported the results of an occupational study was Samuel Milham in 1982.[7] He analyzed the leukemia mortality of white males in Washington State occupationally exposed to electrical and magnetic fields. He observed an excess proportional mortality ratio (PMR) among several trades known to be exposed to EMF: electricians, PMR = 138; power station operators, PMR = 259; aluminum workers, PMR = 189; all electrical occupations, PMR = 137.

Within 6 years, from 1982 to 1988, no less than 12 such communications were published, most of them as short papers or letters to editors.[8-18] As seen in Figure 1, the majority of these reports observed an excess of leukemia among broadly defined "electrical occupations" and this excess was higher for acute myeloid leukemia (Figure 2). Pooled analyses have established a significant excess of all leukemia with a risk estimate of 1.18 (1.09–1.29) and a significant excess of acute myeloid leukemia with a risk estimate of 1.46 (1.27–1.64).[19] These results are for all leukemia and for all workers. More spectacular excesses were noted for specific leukemia types and/or specific occupational groups, but these excesses varied between studies and no findings were clearly consistent.

Everyone, including the authors, recognized that these exploratory studies were gross, numbers were small, exposure was ill defined, statistical analyses often were weak, and no confounders have been controlled for. Consequently, the results can only be indicative and need to be reassessed by using stronger and better study designs.

B. HYPOTHESIS TESTING STUDIES

In an attempt to test the proposed hypothesis, several case-control and cohort studies have been conducted. The case-control studies have addressed mainly two types of cancer, leukemia[20-24] and brain cancer.[23-29] The cohort studies have investigated many occupational groups: telephone operators, electronic industry workers, electrical engineers, telecommunication industry workers, linemen, station operators, electricians, amateur radio operators, telephone company workers, electrical utility workers, and radiomen.[18,30-43]

Their findings are tabulated in Table 1. These studies are quite impressive: they are usually well designed, with reasonably large numbers of cases and they extend over long periods of time. All five leukemia case-control studies have yielded positive findings. In general, excesses have been reported for all leukemia and acute

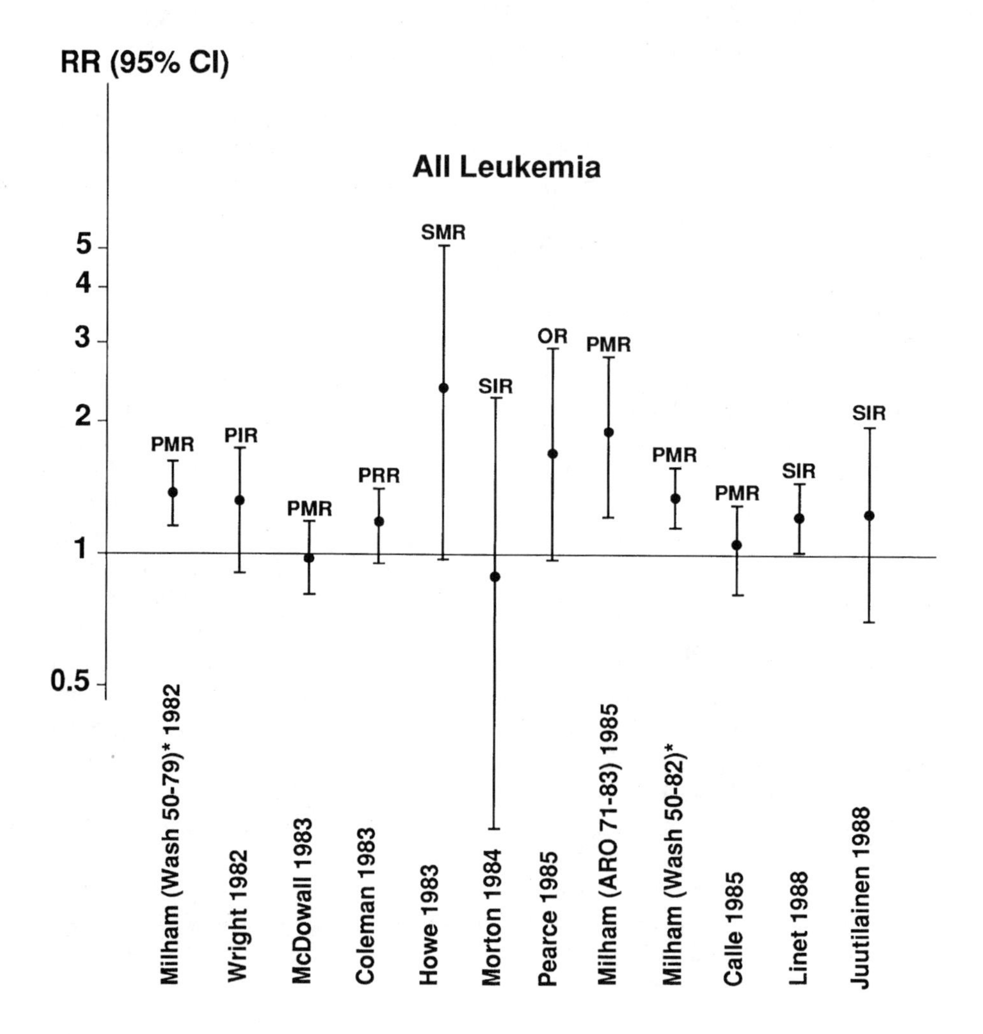

Figure 1 Leukemia risks among electrical worker.

myeloid leukemia. Sometimes, chronic lymphoid leukemia was also elevated. Out of seven brain cancer case-control studies, five have yielded positive results. Some have reported highly elevated odds ratios (13.10 among Texas utility employees).[27] At least three studies indicated the presence of a exposure-response relationship between EMF exposure and brain cancer.

One striking observation from these studies, however, is that whereas most case-control studies have yielded elevated odds ratios among workers believed to be exposed to EMF, cohort studies have yielded almost no such excess of risk. Out of 11 cohort studies of electrical workers, only 3 have shown a significant excess of

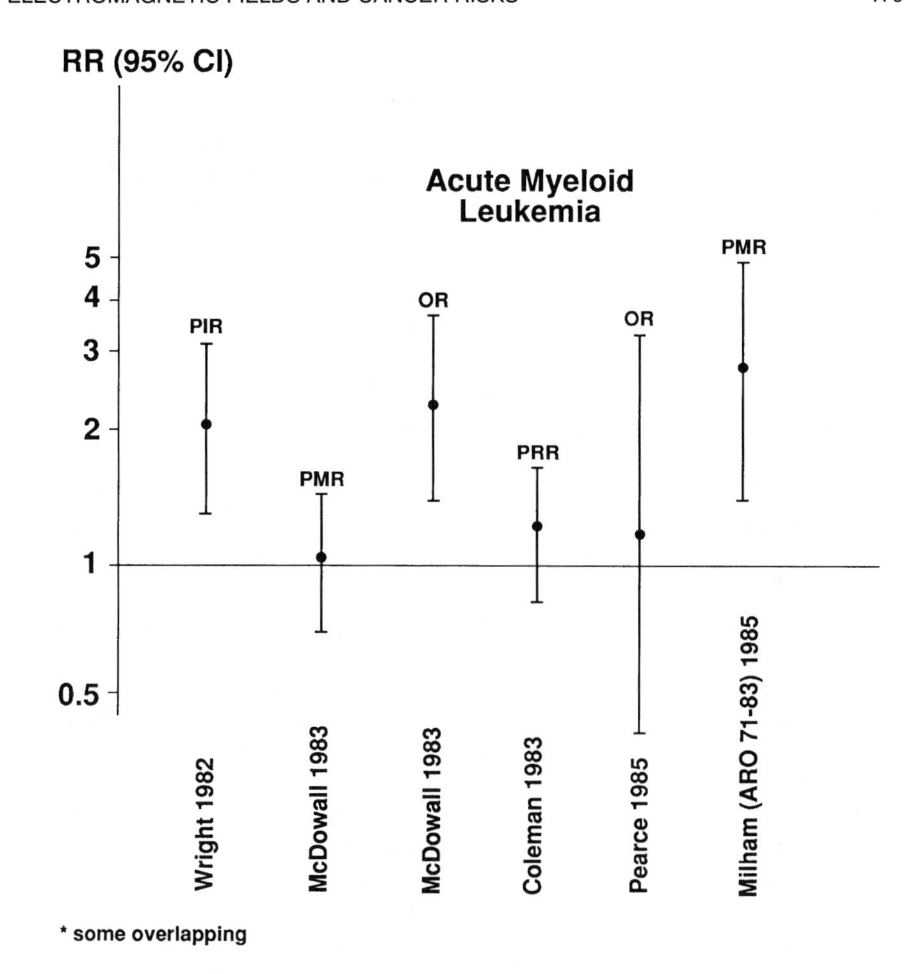

Figure 2 Acute myeloid leukemia risks among electrical worker.

Table 1 Hypothesis Testing Studies

		n Studies	Negative Findings	Questionable Findings	Positive Findings
Leukemia	c/c	5	0	0	5
	co	11	8	0	3
Brain cancer	c/c	7	1	1	5
	co	11	9	1	1
Skin melanoma	c/c	2	1	0	1
	co	7	2	1	4

c/c, case control; co, cohort.

leukemia and 1 an excess of brain cancer. This is rather puzzling. One would have expected cohort studies to yield the same results as the case-control studies.

Two reasons can be proposed to explain these conflicting results. It is either that the number of exposed men in large cohorts is small and their risk is "diluted" by a large proportion of unexposed men, therefore yielding risk ratios that do not reach statistical significance, or the exposure estimates used in case-control studies do not represent EMF exposure. The two explanations are quite possible at this time. However, it is the most recent cohort studies that have yielded significant excess of leukemia, which may indicate that better identification of exposure may confirm the putative association.

III. LEUKEMIA AMONG WELDERS

Welders are exposed to intense electric and magnetic fields. Theoretically, they are among the workers most exposed to EMF. In 1987, Stern[44] conducted an extensive review of 15 cancer studies of welders. He observed that excess lung cancer was reported in most studies but not leukemia. He proposed that this observation goes against the hypothesis of an association of EMF and leukemia.

At approximately the same time, in a case-control study of chronic myeloid leukemia, Preston-Martin and Peters[45] reported a highly elevated odds ratio (OR adjusted 25.4; 2.78–232.54) for welders. This last finding is remarkable in view of the several negative studies previously cited.

What does it mean that welders, who are among the most exposed workers, do not show an excess of leukemia? Why are cohort study results so different from case-control studies? Here again, one can only propose some hypothetical reasons.

Either working under EMF fields does cause cancer and in that context, welders' exposure to EMF would be different from the exposure of electrical workers and cohort studies have yielded negative results just due to the fact that they were not sensitive enough to pick up a real excess among exposed workers, or there is no such effect, in which case the surrogates used for estimating EMF exposure in case-control studies actually measured something else, such as exposure to potential occupational confounders (solvents, creosotes, soldering fumes, etc.). At the moment, it is difficult to say. Much hope is being put on new "third generation" projects. These are large projects with hundreds of thousands of electric utility workers in which exposure is being monitored by careful measurements. They should shed more light on the question.

IV. PROPOSED MECHANISM OF ACTION

When first raised, the carcinogenicity of EMF was believed to be biologically impossible: EMF did not carry enough energy to interact with the cell and to cause changes across the cell membrane. Since then, some effects of EMF on the cell have been observed, such as changes in calcium transfer across the cell membrane and alteration in the DNA synthesis rates and transcription patterns of RNA, with the resultant production of structurally changed proteins. But the most intriguing and worrisome effect noted so far has been the capacity for EMF to diminish the level of circulating melatonin in exposed animals and humans.[46]

Experimental work in laboratories has shown that mean night-time pineal melatonin levels for rats exposed to 39 kV/m electric fields were lower than controls during exposure and returned to normal within 3 d after cessation of exposure. Electric fields of 10, 65, and 130 kV/m lead to a similar response on rats exposed *in utero*. The same phenomenon has been observed in human volunteers.

These observations have led Stevens[47] to propose a mechanism by which EMF may contribute to cancer. According to him, the reduced melatonin production in an exposed individual will generate an increase in estrogen secreted by the ovary and in prolactin secreted by the pituitary gland. It is demonstrated from animal studies that increased estrogen and prolactin contribute to the turnover of breast epithelial stem cells. This increase is known to result in accelerating the growth of hormone-sensitive cancers. It is known that by doing the reverse, meaning increasing melatonin or decreasing estrogen and prolactin, one can delay and even reverse the development of hormone-sensitive cancers.

V. EMF EXPOSURE AND HUMAN BREAST CANCER

If the melatonin theory is correct, one would expect to see an increase in hormone-sensitive cancers among people exposed to EMF. Recently, in a study conducted among telephone workers in the U.S., Matanowski et al.[48] observed two cases of breast cancer among men in a cohort of 9561 office technicians, whereas none would have been expected. The authors indicated that these office technicians were exposed to fields generated by switching on and off electrical devices, and proposed that this may be a clue toward supporting Steven's hypothesized mechanism.

In 1990, Demers et al.[49] looked at the occupational history of 227 male breast cancers and compared them with 300 controls from the general population. This was an attempt to verify the observations of Matanowski and colleagues. They found an elevated odds ratio of 1.8 (1.0–3.2) for all electricity exposed jobs. He found an odds ratio of 6.0 (1.7–21.5) for electricians, telephone linemen, and electrical power workers and an odds ratio of 2.9 (0.8–10.2) for radio and communication workers. "The risk was highest among subjects who were first employed in jobs with exposure before the age of 30 years and who were initially exposed at least 30 years prior to diagnosis".[49]

Also, a study from Norway was reported recently on male breast cancer.[50] It was a cohort study of electrical workers censored in 1960 and 1970. Whereas 5.81 male breast cancers were expected, 12 were observed, for a standardized incidence ratio (SIR) of 207 (107–361). These observations of excess risk of male breast cancer among electrical workers are, to say the least, most intriguing and seem to support the melatonin theory.

If the melatonin theory turns out to be the right one, we should become very alert to the hormone-sensitive cancers, such as breast cancer, prostate cancer, and skin melanoma. This latter one is on the increase in many countries and has been reported in excess in several studies of electrical engineers. These new avenues may guide us into an unforeseen Pandora's box.

VI. CONCLUSION

Research activities to unveil the putative relationship between cancer and exposure to EMF has been ongoing for 12 years. Evidence has always remained unconvincing but it has never faded away. Two observations have been consistently repeated: an association between childhood cancer (leukemia) and wiring configuration around residences and an excess of leukemia and brain cancer among electrical workers. The new theory on melatonin is opening new avenues for more research. Of particular concern is the possibility that EMF could accelerate the growth of hormone-sensitive cancers. With this new knowledge and the new capacity for measuring several parameters of exposure with accuracy, the future looks promising toward the development of a major breakthrough on this issue of electromagnetic fields and cancer risks.

ADDENDUM

Since this chapter was written, two important studies have been reported from Sweden. Feychting and Ahlbom,[51,52] of the Karolinska Institute in Stockholm, carried out a case-control study to investigate residential magnetic field exposure and cancer. Their design took advantage of the population registry systems in Sweden, thereby minimizing the potential for selection bias, a concern in previous studies. They defined the study population as people who lived within a 300-m corridor around 220- and 440-kV power lines from 1960 to 1985, where transmission lines were the dominant source of exposure, thereby providing a unique historical approach to magnetic field exposure assessment. Furthermore, they utilized several techniques to assess exposure: (1) spot meters, (2) correlation between spot meters and line load, and (3) for a sample of the subjects, 24-h measurements. This study consisted of 325 leukemia and 223 brain tumor cases and 1091 controls. The aim of this study was to test the hypothesis that exposure to magnetic fields of the type generated by high-voltage power lines increases cancer incidence. They focused primarily on leukemia and brain tumors, but also included other cancers in children that were diagnosed between 1960 and 1985. Childhood leukemia was reported to have an OR of 2.7 (1.0–6.3) with a trend test of p-value 0.02 for exposures of 0.2 μT and over-using calculated historical fields. The OR increased to 3.8 (1.4–9.3) with trend test of p-value of 0.005 when the upper cutoff was shifted to 0.3 μT, although in adults exposure to magnetic fields of 0.2 μT demonstrated no significant OR for acute myeloid (AML) and chronic myeloid leukemia (CML). No association was seen in brain tumors.

In another study, Floderus et al.[53] focused on 250 male leukemia and 261 male brain tumors diagnosed during 1983 to 1987. These men were 20 to 64 years old as of 1980 and residents of Mid Sweden or Gothenburg County. The control group consisted of 1121 men from the 1980 census matched by age. Exposure information was collected using three EMDEX-100 and six EMDEX-C meters produced by the Electrical Power Research Institute (EPRI) and Electric Field Measurements (EFM).

Along with the dosimeter they used a computer program (DATACALC) for an analysis of the single measurement. The objective of the study was to investigate the possible association between extremely low frequency electromagnetic fields and cancer. They found that the OR for chronic lymphocytic leukemia (CLL) was 3.0 (1.6–5.8) in the highest quartile (0.28 μT). One of the most important findings of this study was the fact that there was an increased risk with increasing level of exposure (i.e., dose-response).

It is interesting to note that these studies utilized more sensitive indicators of exposure and appear to show some significant trends for childhood leukemia.

REFERENCES

1. Wertheimer, N. and Leeper, E., Electrical wiring configurations and childhood cancer, *Am. J. Epidemiol.*, 109, 273, 1979.
2. Myers, A., Cartwright, R.A., Bonnell, J.A., Male, C.R., and Cartwright, S.C., Overhead power lines and childhood cancer, Presented at the International Conference on Electric and Magnetic Fields in Medicine and Biology, London, 1985.
3. Tomenius, L., 50-Hz electromagnetic environment and the incidence of childhood tumours in Stockholm country, *Bioelectromagnetics*, 7, 191, 1986.
4. Savitz, D.A., Case-Control Study of Childhood Cancer and Residential Exposure to Electric and Magnetic Fields. Report to the New York State Department of Health, Power Lines Project, 1987.
5. Fulton, J.P., Cobb, S., Preble, L., Leone, L., and Forman, E., Electrical wiring configuration and childhood leukemia in Rhode Island, *Am. J. Epidemiol.*, 111, 292, 1980.
6. London, S.J., Thomas, D.C., Bowman, J.D., Sobel, E., Cheng, T., and Peters, J.M., Exposure to residential electric and magnetic fields and risk of childhood leukemia, *Am. J. Epidemiol.*, 134, 923, 1991.
7. Milham, S., Mortality from leukemia in workers exposed to electrical and magnetic fields, *N. Engl. J. Med.*, 307, 249, 1982.
8. Wright, W.E., Peters, J.M., and Mack, T.M., Leukemia in workers exposed to electrical and magnetic fields, *Lancet*, 2, 1160, 1982.
9. McDowall, M.E., Leukemia mortality in electrical workers in England and Wales, *Lancet*, 1, 246, 1983.
10. Coleman, M., Bell, J., and Skeet, R., Leukemia incidence in electrical workers, *Lancet*, 1, 982, 1983.
11. Howe, G.R. and Lindsay, J.P., A follow-up study of a ten-percent sample of the Canadian Labour Force. 1. Cancer mortality in males, 1965-73, *J. Natl. Cancer Inst.*, 70, 37, 1983.
12. Morton, W. and Marjanovic, D., Leukemia incidence by occupation in Portland-Vancouver Metropolitan area, *Am. J. Ind. Med.*, 6, 185, 1984.
13. Pearce, N.E., Sheppart, A.R., Howard, J.K., Fraser, J., and Lilley, B.M., Leukemia in electrical workers in New Zealand, *Lancet*, 1, 811, 1985.
14. Milham, S., Silent keys: leukemia mortality in amateur radio operators, *Lancet*, 1, 811, 1985.
15. Milham, S., Mortality in workers exposed to electromagnetic fields, *Environ. Health Perspect.*, 62, 297, 1985

16. Calle, E. and Savitz, D.A., Leukemia in occupational groups with presumed exposure to electrical and magnetic fields, *N. Engl. J. Med.*, 313, 1476, 1985.
17. Linet, M., Malker, H., McLaughlin, J. et al., Leukemias and occupation in Sweden: a registry-based analysis, *Am. J. Ind. Med.*, 14, 319, 1988.
18. Juutilainen, J., Pukkala, E., and Laara, E., Results of epidemiological cancer study among electrical workers in Finland, *J. Bioelectricity*, 7, 119, 1988.
19. Coleman, M. and Beral, V., A review of epidemiological studies of the health effects of living near or working with electricity generation and transmission equipment, *Int. J. Epidemiol.*, 17, 1, 1988.
20. Gilman, P.A., Ames, R.G., and McCawley, A., Leukemia risk among U.S. white male coal miners, *J. Occup. Med.*, 27, 669, 1985.
21. Stern, F.B., Waxweiler, R.A., Beaumont, J.J. et al., A case-control study of leukemia at a naval nuclear shipyard, *Am. J. Epidemiol.*, 123, 980, 1986.
22. Flodin, U., Fredriksson, M., Axelson, O., Persson, B., and Hardell, L., Background radiation, electrical work, and some other exposures associated with acute myeloid leukemia in a case-referent study, *Arch. Environ. Health*, 41, 77, 1986.
23. Coggon, D., Pannett, B., Osmond, C., and Acheson, E.D., A survey of cancer and occupation in young and middle aged men. II. Non-respiratory cancers, *Br. J. Ind. Med.*, 43, 381, 1986.
24. Pearce, N., Reif, J., and Fraser, J., Case-control studies of cancer in New Zealand electrical workers, *Ind. J. Epidemiol.*, 18, 55, 1989.
25. Lin, R.S., Dischinger, P.C., Conde, J., and Farrell, K.P., Occupational exposure to electromagnetic fields and the occurrence of brain tumours: an analysis of possible associations, *J. Occup. Med.*, 27, 413, 1985.
26. Thomas, T.L., Stolley, P.D., Sternhagen, A. et al., Brain tumour mortality among men with electrical and electronics jobs: a case-control study, *J. Natl. Cancer Inst.*, 79, 233, 1987.
27. Speers, M.A., Dobbins, J.G., and Miller, V.S., Occupational exposures and brain cancer mortality: a preliminary study of East Texas residents, *Am. J. Ind. Med.*, 13, 629, 1988.
28. Savitz, D.A., John, E.M., and Kleckner, R.C., Magnetic field exposure from electric appliances and childhood cancer, *Am. J. Epidemiol.*, 131, 763, 1990.
29. Brownson, R.C., Reif, J.S., Change, J.C., and Davis, J.R., An analysis of occupational risks for brain cancer, *Am. J. Public Health*, 80, 169, 1990.
30. Wilklund, K., Einhorn, J., and Eklund, G., An application of the Swedish Cancer Environment Registry. Leukemia among telephone operators at the telecommunications administration in Sweden, *Int. J. Epidemiol.*, 10, 373, 1981.
31. Vagero, D. and Olin, R., Incidence of cancer in the electronics industry: using the new Swedish Cancer Environment Registry as a screening instrument, *Br. J. Ind. Med.*, 40, 188, 1983.
32. Olin, R., Vagero, D., and Ahlbom, A., Mortality experience of electrical engineers, *Br. J. Ind. Med.*, 42, 211, 1985.
33. Vagero, D., Ahlbom, A., Olin, R., and Sahlsten, S., Cancer morbidity among workers in the telecommunications industry, *Br. J. Ind. Med.*, 42, 191, 1985.
34. Tornqvist, S., Norell, S., Ahlbom, A., and Knave, B., Cancer in the electric power industry, *Br. J. Ind. Med.*, 43, 212, 1986.
35. McLaughlin, J.K., Malker, H.S.R., Blot, W.J. et al., Occupational risks for intracranial gliomas in Sweden, *J. Natl. Cancer Inst.*, 78, 253, 1987.
36. Milham, S., Increased mortality in amateur radio operators due to lymphatic and haematopoietic malignancies, *Am. J. Epidemiol.*, 127, 50, 1988.

37. Guberan, E., Usel, M., Raymong, L., Tissot, R., and Sweetman, P.M., Disability, mortality, and incidence of cancer among Geneva painters and electricians, *Br. J. Ind. Med.*, 46, 16, 1989.
38. DeGuire, L., Theriault, G., Iturra, H., Provencher, S., Cyr, D., and Case, B.W., Increased incidence of malignant melanoma of the skin in workers in a telecommunications industry, *Br. J. Ind. Med.*, 45, 824, 1988.
39. Lin, R.S., Cancer Mortality Patterns Among Employees of Telecommunication Industry in Taiwan, Presented at the Annual Review of Research on Biological Effects of 50 and 60 Hz Electric and Magnetic Fields, U.S. Department of Energy and EPRI, Portland, Oregon, November, 1989.
40. Matanowski, G., The Hopkins Telephone Worker Study, *Health Safety Rep.*, 7, 3, 1989.
41. Koifman, S., Echenique, I.B., Farias, A.M., Monteiro, G., and Koifman, R.J., Cancer mortality among workers in an electrical company in Rio De Janeiro, Brazil. Environmental Stress, Proceedings of the First Symposium on Environmental Stressors, Tampere, Finland, July, 1989, 225, 19.
42. Garland, F.C., Shaw, E., Gorham, E.D., Garland, D.F., White, M.R., and Sinsheimer, P., Incidence of leukemia in occupations with potential electromagnetic fields exposure in United States Navy personnel, *Am. J. Epidemiol.*, 132, 293, 1990.
43. Vagero, D., Swerdlow, A.J., and Beral, V., Occupation and malignant melanoma: a study based on cancer registration data in England and Wales and in Sweden, *Br. J. Ind. Med.*, 47, 317, 1990.
44. Stern, R.M., Cancer incidence among welders: possible effects of exposure to extremely low frequency electromagnetic radiation (ELF) and to welding fumes, *Environ. Health Perspect.*, 76, 221, 1987.
45. Preston-Martin, S. and Peters, J.M., Prior employment as a welder associated with the development of chronic myeloid leukemia, *Br. J. Cancer*, 58, 105, 1988.
46. Wilson, B.W. and Anderson, L.E., ELF electromagnetic field effects on the pineal gland in extremely low frequency electromagnetic fields, in *The Question of Cancer*, Wilson, B.W., Stevens, R.G., and Anderson, L.A., Eds., Battelle Press, Columbus, 1990.
47. Stevens, R.G., Electric power use and breast cancer: a hypothesis, *Am. J. Epidemiol.*, 125, 556, 1987.
48. Matanowski, G.B.M., Breysse, P.N., and Elliott, E.A., Electromagnetic field exposure and male breast cancer, *Lancet*, 337, 737, 1991.
49. Demers, P.A., Thomas, D.B., Rosenblatt, K.A. et al., Occupational exposure to electromagnetic radiation and breast cancer in males, *Am. J. Epidemiol.*, 134, 340, 1991.
50. Tynes, J.T. and Andersen, A., Electromagnetic fields and male breast cancer, *Lancet*, 336, 1596, 1990.
51. Feychting, M. and Ahlbom, A., Magnetic fields and cancer in people residing near Swedish high voltage power lines, *Am. J. Epidemiol.*, 138, 467, 1993.
52. Feychting, M. and Ahlbom, A., Megnetic fields, leukemia, and central nervous system tumors in Swedish adults residing near high-voltage power lines, *Epidemiology*, 5, 501, 1994.
53. Floderus, B., Persson, T., Stenlund, C. et al., Occupational exposure to electromagnetic fields in relation to leukemia and brain tumours. A case-control study in Sweden, Cancer causes and control, 4, 465, 1993.

CHAPTER 12

Recent Progress in Childhood Lead Exposure

Herbert L. Needleman

I. INTRODUCTION

Remarkable progress in understanding the pathophysiology, epidemiology, and long-term consequences of childhood lead exposure has been made in the last decade. In response, the federal government has pronounced lead exposure as the "most serious environmental disease of American children", and accordingly shifted its strategy from case finding and treatment to authentic primary prevention.

In addition to discussing recent progress in the epidemiology of childhood lead exposure, this chapter examines steps to eliminate the problem permanently. It also investigates some of the forces that have retarded progress in prevention. Unless these forces are understood, the strong promise that this disease will be banished to the medical history texts will be forfeited.

In the second century B.C., Dioscorides, perhaps the first neurotoxicologist, said that "Lead makes the mind give way".[1] Two hundred years ago, Benjamin Franklin wrote of the "dry gripes" (colic) and "dangles" (wrist drop) that affected typesetters and painters.[2] Almost a century ago, childhood lead poisoning was first described by a pediatric house officer, A.J. Turner, at the Brisbane Children's Hospital.[3] His colleague, J. Lockhart Gibson, had been referred a number of cases of ophthalmoplegia and retinitis. He also diagnosed these as plumbism. Gibson and Turner, by painstaking observation, noticing that many children were admitted shortly after they had moved to a different house, concluded that the disease was one of "habitation".

In 1904, Gibson published "A Plea for Painted Railings and Painted Walls of Rooms as the Source of Lead Poisoning among Queensland Children".[4] This was greeted by considerable derision by both medical and commercial interests, but in 1920 the Australian Medical Congress passed a resolution asking for a ban on lead paint in houses. Fifty years later, the U.S. followed suit.

0-87371-573-X/95/$0.00+$.50

Childhood lead poisoning was described in the U.S. in the early part of the 20th century.[5,6] It was generally believed that there were only two possible outcomes to a case of acute lead toxicity, death or complete recovery. Byers, a pediatric neurologist, followed up 20 children assumed to be completely recovered from plumbism, and found that 19 had school or behavioral problems.[7] He raised for the first time the question of whether lead toxicity that had not been diagnosed was a cause of school failure. This paper opened the modern era of lead toxicology.

II. STUDIES OF ASYMPTOMATIC EXPOSURE

It was assumed in the 1940s and 1950s that lead toxicity occurred only in the presence of symptoms of headache, clumsiness, constipation, or vomiting. Perlstein and Attala[8] followed up recovered children and found that 39% of those who recovered had sequelae. Of children with elevated blood lead levels but no presenting symptoms, 9% were mentally retarded.

The study of "silent" lead poisoning began in the 1970s. Early studies were inconsistent in conclusions. In 1979, my group identified four major design problems in previous published studies of "low level lead exposure": poor markers of exposure, weak measures of outcome, inadequate control of covariates, and selection bias. Studies that classified children by blood lead level could misclassify subjects, since the blood is a short-term storage system. Group or screening tests are inadequate to measure small changes in function at lesser doses. Many factors affect development, and some could be correlated with lead, and thus confound its effects. Finally, subjects drawn from psychiatric or pediatric clinics may not be representative of the population of children at risk. We designed a study to address these issues. We used as a marker of exposure, dentine lead levels, since these represent past exposure to lead.[9] We gave our subjects a large battery of measures tapping psychometric intelligence, attention, language function, and behavior, and controlled for a number of nonlead covariates. Our subjects were chosen from ordinary first and second grades in Somerville and Chelsea, Massachusetts.

After control of covariates, high lead children had six points lower mean IQ score than their low lead counterparts. They also had inferior scores on measures of speech and language function, and on attention. When teachers rated children on an 11-item classroom behavior scale, a dose-dependent relationship between lead and outcome was found. Similar results on the rating scale were subsequently reported by Yule et al.[10] and Hatzakis et al.[11]

Although the mean difference in IQ scores was six points, a shift of the distributions of this magnitude increases the proportion of children with severe deficits fourfold, from 4 to 16% (Figure 1). In addition, this shift truncates the top of the distribution, and prevents 5% of the sample from achieving superior function (IQ >125). The Office of Technology Assessment has estimated the number of American children who were deprived of truly superior function because of this shift in function due to low-level lead exposure at 1.3 million.[12]

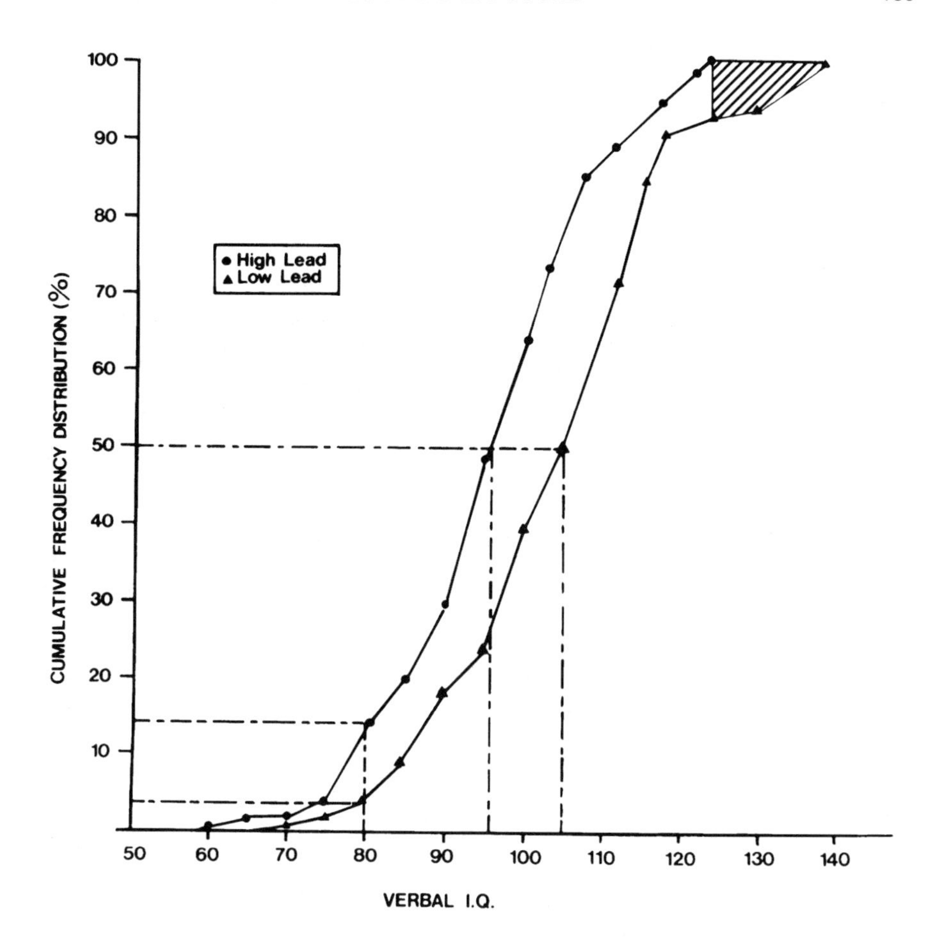

Figure 1 Cumulative frequency of IQ scores in high and low lead subjects. A shift in the median IQ of 6 points is associated with a fourfold increase in severe deficits (IQ < 80).

Among the most noteworthy studies of low-level lead toxicity published in the 1980s were those of Fulton et al.,[13] examining middle class children in Edinburgh, Scotland; Hatzakis et al.,[14] in Lavrion, Greece; and Hansen et al[15] in Denmark. These investigations showed a deficit in mean IQ in the range of four to seven points. The exposure levels in the Danish and Scottish studies were considerably lower than those previously reported, with changes observed at blood lead levels as low as 10 μg/dl. In Greece, the changes were observed at 25 μg/dl. The Scottish and Greek studies also showed impaired attention associated with lead, and the Danish studies showed a lead-associated increase in learning disabilities.

III. BEHAVIORAL EFFECTS

It had long been observed that children who had been lead poisoned frequently were overactive and aggressive. Teachers in the 1979 Boston area study, who did not

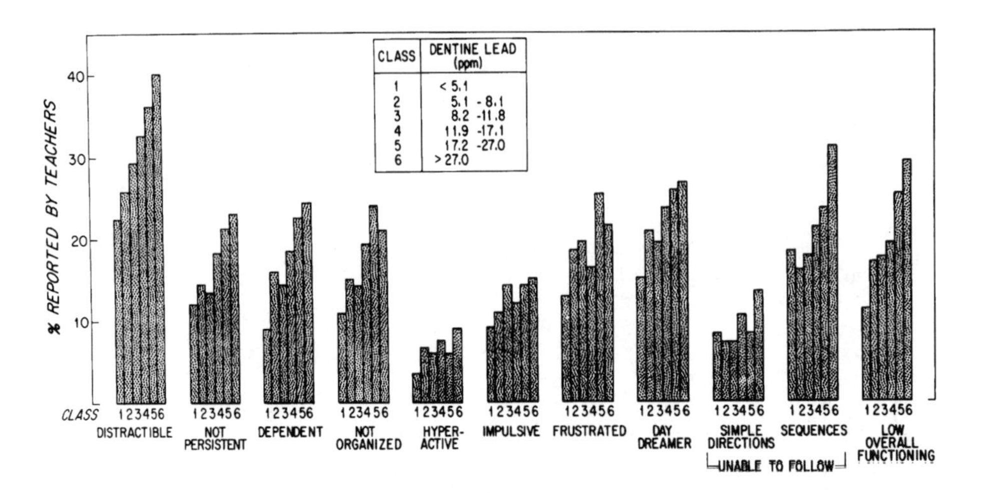

Figure 2 Teachers' ratings of eleven classroom behaviors in relation to tooth lead levels. Teachers' knew the subjects at least 2 months, but were blind to their lead levels. Dentine lead levels are arranged in ascending order, and plotted against the rate of bad evaluations for each item.

know their pupils' lead levels, reliably reported that children with higher lead levels were more distractible, disorganized, less able to follow directions, and were more hyperactive.

Lansdown and colleagues found that lead was associated with higher scores on the overactivity factor of the Rutter B2 scale, and on the conduct problem, inattentive, passive, and hyperactive scales of the Connors Questionnaire and the Tailor-Sandberg Hyperactivity Score. Similar relationships to those shown in Figure 2 were also seen. Raab et al., studying the same subjects as Fulton,[16] found that lead was related to inspection time, a measure of attention. Thomson et al.[17] reported that lead was related to hyperactivity and aggressive/antisocial scores on the Rutter Teachers' Scale. Similarly, Fergusson and colleagues showed that lead was significantly related to scores on the inattentive/restlessness items of the Rutter and Connors Scales after controlling for covariates.[15] These reports suggest that attention function may be a sensitive target for lead, and that a portion of children with attention deficit disorder (ADD) may have had unrecognized lead toxicity.

IV. LONG-TERM CONSEQUENCES OF EARLY EXPOSURE

We followed up on our subjects at two time periods, when they were fifth graders and when they were young adults. In the fifth grade, lead was significantly associated with increased grade retention, and high lead subjects tended to have lower IQ scores and greater need for special school services.[19]

Among young adults, the results were even more striking. We were able to bring 132 of our original sample of 270 subjects back for testing 12 years later.[20] Their mean age was 18 years. Having elevated tooth lead was associated with a number of problems in adulthood. High lead subjects had an adjusted risk of 6.8 of not graduating from high school, and a risk of 5.8 of reading disability. Of the 10 subjects who had a diagnosis of clinical plumbism in 1976, 50% had a reading disorder and 42% had school failure. Lead exposure was also associated with lower class standing in the final year of high school, with increased absenteeism, lower vocabulary scores, and impaired fine motor function. Lead exposure at these doses seems to produce permanent effects that influence academic adjustment.

V. PRENATAL EXPOSURE TO LEAD

Lead crosses the placenta, and can be measured in the umbilical cord blood.[21] We studied the effects of exposure in a cohort of 12,000 subjects enrolled at the Boston Hospital for Women. Umbilical cord blood lead levels declined approximately 10% per year. This was closely correlated with the reduction of lead in gasoline. We first examined the relationship between umbilical cord lead and neonatal status in 5000 births. Controlling for a number of factors, lead was significantly associated with the rate of minor malformations. Others have reported no relationship between cord blood lead and malformations, but their sample sizes were too small to permit that assertion.[22] Other investigations have found that lead exposure is related to intrauterine growth and growth during the first year of life.[23]

We followed a cohort of 249 children forward in time at 1-, 6-, 12-, 24-, and 57-month intervals, and again at 10 years of age. At 24 months, umbilical cord blood lead was significantly related to Bayley Developmental Scales, after covariate adjustment.[24] At 57 months of age, the prenatal exposure was no longer influential, except in one group of subjects: infants of low SES who had postnatal exposure. The effect of 24-month blood lead levels on intelligence was significant. The high lead group had a decrease in IQ of eight points. It is notable that the lower boundary of the high lead group was 10 mg/dl.[25] Other investigators who found an effect of early exposure on outcome were Dietrich et al.[26] and McMichael and colleagues.[27]

VI. SYNTHESIS OF STUDIES

The question of the impact of lead on IQ at lesser doses may have engendered more controversy than any other issue in environmental health. There are a number of reasons for this. The effects of lead are nonspecific. Many factors can lower IQ scores, increase restlessness and distractibility, and impair speech development. Many of these, such as undernutrition, iron deficiency, infections, head injury, and inferior prenatal care, are more frequent in the poor.

Not all studies have shown an effect, and some argue that this inconsistency means that there is no effect in nature.

A recent meta-analysis of all 24 informative modern studies handled those using blood separately from those using teeth.[28] A separate and more detailed analysis was done for the 12 studies that contained enough information to estimate effect size, expressed as partial r. The values for partial r ranged from –0.27 to –0.003. The joint P value for all blood lead studies was <0.0001. For the tooth studies the P value was <0.0005. When all 24 studies were included in the analysis, the P value was <0.0001. We then investigated whether this observation could be due to the differential publication of studies that showed a positive effect. A "file drawer" analysis was done. This procedure estimates how many unpublished no-effect studies would be required to vitiate the observed P value. To bring the P value for this study to nonsignificance, there would have to be 93 unpublished studies that showed no lead effect in file drawers around the world. Since this paper was published, there have been at least five other reports showing an effect of lead at low dose on IQ.

VII. WHAT IS THE TOXIC DOSE OF LEAD?

The defined toxic dose of lead has decreased over time as studies with larger samples, more sensitive measures of outcome, and better statistical techniques have been reported. In the 1960s, the toxic level was defined as 60 μg/dl. This was revised downward by the Department of Health, Education and Welfare (DHEW) in 1972 on the basis of available information, to 40 μg/dl. The Centers for Disease Control (CDC) reevaluated the literature in 1978, and lowered the threshold to 30 μg/dl, and in 1985 to 25 μg/dl. A thorough reevaluation was undertaken by CDC in 1991. The CDC report concluded:

> New data indicate significant adverse effects of lead exposure in children at blood lead levels previously believed to be safe. Some adverse health effects have been documented at blood lead levels at least as low as 10 micrograms per deciliter of whole blood (μg/dl). The 1985 intervention level of 25 μg/dl is, therefore, being revised downwards to 10 μg/dl.

VIII. WHAT IS THE EXTENT OF EXCESS LEAD EXPOSURE?

The most current data on blood lead levels derive from the National Health and Examination Survey of 1976 to 1980. This is a balanced probability sample of 26,000 American adults and children. A blood lead level was obtained from approximately 5000 children from 9 months to 6 years of age. Since the time this survey was conducted, blood lead levels in this country have declined. A good portion of this decline can be assigned to the removal of lead from gasoline during the conduct of the study.

Mushak and Croccetti reviewed the epidemiology of lead for the Report to Congress of the Agency for Toxic Substances and Disease Registry,[29] and adjusted the prevalences for the secular decline of blood lead concentrations, and estimated the contemporary blood lead concentrations (Table 1). They state:

Table 1 Projected Percentages of Children 0.5–5 Years Old Estimated to Exceed Blood Lead of 15 μg/dl (1984)

Family Income/Race	>15 μg/dl
<$6,000	
White	25.7%
Black	55.5%
>$15,000	
White	7.1%
Black	25.6%

Adapted from the Agency for Toxic Substances and Disease Registry, *The Nature and Extent of Lead Poisoning in Children in the United States*: A Report to Congress, Atlanta, Department of Health and Human Servies, 1988.

In short, about 2.4 million white and black children (ages 6 months to 5 years) or about 17% of such children in U.S. SMSAs, are exposed to environmental sources of lead at concentrations that place them at risk of adverse health effects. This number approaches 3 million black and white children if extended to the entire U.S. child population. If the remaining racial categories are included in these totals, between 3 and 4 million U.S. children may be affected.

IX. WHY HAS PROGRESS IN LEAD CONTROL BEEN SLOW?

There are a number of reasons for exceedingly slow progress in lead poisoning control. Lead produces no characteristic picture, and the effects at lesser doses are not dramatic. Lead exposure has been traditionally considered a disease of poor minority children. Associated with this conviction is the belief that if the mother had cared for the child more assiduously, the problem would not have happened. Once the burden has shifted to the parents, medical and public health responsibility can rest. If lead poisoning were a disease that preferentially sought out middle class children, it would probably have been eradicated by now. Academic pediatrics has to a considerable extent ignored the problem of low-level lead exposure.

There are a number of excellent pediatric departments that give little attention to teaching about low-dose lead toxicity, and some that do not screen for lead in their outpatient clinics. The Department of Housing and Urban Development has not fulfilled its major responsibility in dealing with lead paint in either its public housing or in private housing under its aegis.

The lead industry, through its trade organizations and academic consultants, has worked hard at camouflaging the effects of lead at low dose. This activity is at least 50 years old. When Byers published his paper in 1943, he was threatened with a million-dollar law suit.

There is a widely held misperception that since lead has been removed from gasoline, and a ban on lead in household paint has been in place since 1976, the problem has been taken care of. In fact, there are over 3 million homes in which children live that have deteriorated leaded surfaces. Almost every child in each of these houses has an elevated lead level. Curiously, coexistent with the belief that the problem has been eradicated is the belief that the problem is too big for our society to handle in a time of budgetary deficits.

X. STEPS TOWARDS PRIMARY PREVENTION

In the winter of 1990, Dr. James Mason, Assistant Secretary of Health, directed Dr. Vernon Houk, Director of the Center for Environmental Health and Injury Control of the Centers for Disease Control, to draft a plan to eradicate childhood lead poisoning. On February 21, 1991, the plan was announced. This represented a fundamental shift in the Public Health Service's response to the modern picture of lead toxicity in the U.S.

Instead of finding cases and treating them, the plan called for a staged attack on the problem, consisting of increased childhood lead poisoning prevention programs and activities, effective abatement of leaded homes, reduction of lead from other sources, and a national surveillance effort. Perhaps the most important part of this historic document is the cost benefit analysis that CDC commissioned. The economists monetized the costs for health care, special education, lowered earning rates, and abating a small house. From these data, they estimated that the cost for deleading a modest pre-1950 house at $2225. They estimated the monetized benefits from decreased medical and special educational costs, and from increased wages associated with higher IQs at $4323 per abated unit. The discounted costs for abating all pre-1950 houses was $33 billion. The discounted benefit was $62 billion. CDC's economists tested various models and found that the benefits were robust to changes in assumptions of the costs of abatement, efficacy of abatement, and the discount rate. Because metrics for good health and better living quality are less precise than metrics for labor and material costs, cost benefit analyses generally underestimate the benefit of environmental control. Nevertheless, this analysis provides a sound basis on which to undertake the program. Another unmeasured benefit is the lift to communities that accompanies attention to the quality of the residents' lives, and perhaps most important, the possibility that jobs would be created in these areas.

XI. ENDING CHILDHOOD LEAD POISONING

Lead poisoning, like automobile accidents, violent deaths, and lung cancer, is primarily due to conscious choices in human activity. This should be a source of some optimism. Being manmade, plumbism is potentially completely preventable. A common response is that this is a utopian delusion. Where could the money be found to delead 30 million houses?

The prospect of driving a disease rate down to zero is not new. In 1963, the Executive Secretary of the World Health Organization was approached with a proposal to eradicate smallpox in the world. He thought this was also a utopian vision, and reasoned that since it would fail, it were best if an American was leader of the project. Thus, Dr. Donald Henderson was appointed to direct the smallpox campaign. In 1965 the last case of smallpox was reported. Henderson is now science advisor to the President. Lead poisoning can have the same history given the same vision, commitment, and quality of leadership.

REFERENCES

1. Major, R. H., Some landmarks in the history of lead poisoning, *Ann. Med. History*, 3, 218, 1931.
2. Franklin, B., Pernicious effects of lead, in *The Works of Ben Franklin*, Franklin, B., Ed., B. C. Buzby, Philadelphia, 1818.
3. Gibson, J. L., A plea for painted railings and painted walls of rooms as the source of lead poisoning among Queensland children, *Aust. Med. Gazette*, 23, 149, 1904.
4. Turner, A. J., Lead poisoning among Queensland children, *Aust. Med. Gazette*, 16, 475, 1897.
5. McKhann, C. F. and Vogt, E. C., Lead poisoning in children: with notes on therapy, *Am. J. Dis. Child.*, 32, 386, 1926.
6. Blackfan, K. D., Lead poisoning in children with special reference to lead as a cause of convulsions, *Am. J. Med. Sci.*, 153, 877, 1917.
7. Byers, R. K. and Lord, E. E., Late effects of lead poisoning on mental development, *Am. J. Dis. Child.*, 66, 471, 1943.
8. Perlstein, M. A. and Attala, R., Neurologic sequelae of plumbism in children, *Clin. Pediatr.*, 5, 292, 1966.
9. Needleman, H. L., Gunnoe, C., Leviton, A., Peresie, H., Maher, C., and Barret, P., Deficits in psychological and classroom performance of children with elevated dentine lead levels, *N. Engl. J. Med.*, 300, 689, 1979.
10. Yule, W., Lansdown, R., Millar, I. B. et al., The relationship between blood lead concentrations, intelligence and attainment in a school population: a pilot study, *Dev. Med. Child. Neurol.*, 23, 567, 1981.
11. Hatzakis, A., Kokkevi, A., Katsouyanni, K. et al., Psychometric intelligence and attentional performance deficits in lead-exposed children, in *International Conference: Heavy Metals in the Environment*, Vol. 1, CEP Consultants, Edinburgh, 1987, 204.
12. U.S. Congress, Office of Technology Assessment, *Neurotoxicity, Identifying and Controlling Poisons of the Nervous System*, OTA-BA-436, U.S. Government Printing Office, Washington, D.C., 1990.
13. Fulton, M., Raab, G., Thomson, G., Laxen, D., Hunter, R., and Hepburn, W., Influence of blood lead on the ability and attainment of children in Edinburgh, *Lancet*, 1, 1221, 1987.
14. Hatzakis, A., Kokkevi, A., Maravelias, C. et al., Psychometric intelligence deficits in lead-exposed children, in *Lead Exposure and Child Development: an International Assessment*, 1st ed., Smith, M. A., Grant, L. D., and Sors, A., Eds., Kluwer Academic, London, 1989, 211.
15. Hansen, O. N., Trillingsgaard, A., Beese, I. et al., A neuropsychological study of children with elevated dentine lead level: assessment of the effect of lead in different socio-economic groups, *Neurotoxicol. Teratol.*, 1, 205, 1989.

16. Raab, G. M., Thomson, G. O. B., Boyd, L., Fulton, M., and Laxen, D. P. H., Blood lead levels, reaction time, inspection time and ability in Edinburgh children, *Br. J. Dev. Psychol.*, 8, 101, 1990.
17. Thomson, G. O. B., Raab, G. M., Hepburn, W. S., Hunter, R., Fulton, M., and Laxen, D. P. H., Blood lead levels and children's behavior — results from the Edinburgh lead study, *J. Child. Psychol. Psychiatry*, 30, 515, 1989.
18. Fergusson, D. M., Fergusson, J. E., Harwood, L. J., and Kinzett, N. G., A longitudinal study of dentine lead levels, intelligence, school performance and behavior. III. Dentine lead levels and cognitive ability, *J. Child. Psychol. Psychiatry*, 29, 973, 1988.
19. Bellinger, D., Needleman, H. L., Bromfield, R., and Mintz, M., A follow-up study of the academic attainment and classroom behavior of children with elevated dentine lead levels, *Biol. Trace Elem. Res.*, 6, 207, 1984.
20. Needleman, H. L., Schell, A., Bellinger, D., Leviton, A., and Allred, E. N., The long-term effects of exposure to low doses of lead in childhood: an 11-year follow-up report, *N. Engl. J. Med.*, 322, 83, 1990.
21. Scanlon, J., Umbilical cord blood lead concentration: relationship to urban or suburban residency during gestation, *Am. J. Dis. Child.*, 121, 325, 1971.
22. Ernhart, C. B., Wolf, A. W., Kennard, J. J. et al., Intrauterine exposure to lead: the status of the neonate, *Arch. Environ. Health*, 41, 287, 1986.
23. Shukla, R., Bornschein, R. L., Dietrich, K. N. et al., Effects of fetal and early postnatal lead exposure on child's growth and stature: the Cincinnati lead study, in *Sixth International Conference on Heavy Metals in the Environment*, Vol. 1, Lindberg, S. and Hutchinson, T., Eds., CEP Consultants, Edinburgh, 1987, 210.
24. Bellinger, D., Leviton, A., Waternaux, C. et al., Longitudinal analyses of prenatal and postnatal lead exposure and early cognitive development, *N. Engl. J. Med.*, 316, 1037, 1987.
25. Bellinger, D., Stiles, K. M., and Needleman, H. L., Low-level lead exposure, intelligence, and academic achievement: a long-term follow-up study, *Pediatrics*, 90, 855, 1992.
26. Dietrich, K., Krafft, K., Beir, M., and Bornschein, R., Early effects of fetal lead exposure: developmental findings at six months, in *Heavy Metals in the Environment: International Conference, New Orleans. 1987*, Lindberg, S. and Hutchinson, T., Eds., CEP Consultants, Edinburgh, 1987.
27. McMichael, A. J., Baghurst, P. A., Wigg, N. R., Vimpani, G. V., Robertson, E. F., and Roberts, R. J., Port Pirie cohort study: environmental exposure to lead and children's abilitites at the age of four years, *N. Engl. J. Med.*, 319, 468, 1988.
28. Needleman, H. L. and Gatsonis, C., Low level lead exposure and the IQ of children, *JAMA*, 263, 673, 1990.
29. Agency for Toxic Substances and Disease Registry, The Nature and Extent of Lead Poisoning in Children in the United States: A Report to Congress, Atlanta, Department of Health and Human Services, 1988.

CHAPTER 13

Sensitive Subgroups and Normal Variation in Pulmonary Function Response to Air Pollution Episodes

Bert Brunekreef, Patrick L. Kinney, James H. Ware, Douglas Dockery, Frank E. Speizer, John D. Spengler, and Benjamin G. Ferris, Jr.

I. INTRODUCTION

A growing literature indicates that children experience short-term declines in pulmonary function level during and shortly after episodes of high outdoor air pollution.[1-7] The evidence has been gathered by performing repeated pulmonary function measurements in cohorts of children exposed to episodes of elevated air pollution to determine whether pulmonary function levels vary inversely with air pollutant concentration. Such studies produce a natural measure of response for each participant: the coefficient of regression of the child's pulmonary function measurements on the air pollution concentrations on the day or days preceding the examinations.

Stebbings and Fogelman[1] studied 224 children over a period of 6 d immediately after an episode of high concentrations of total suspended particulates (TSP) and sulfur dioxide (SO_2). They found that pulmonary function was depressed during the air pollution episode. Dockery et al.[2] obtained pulmonary function measurements at weekly intervals for 6 to 8 weeks in four different studies involving a total of 331 children. These examinations spanned several episodes of high TSP and SO_2 concentrations. In these studies, the majority of children has negative regression slopes of pulmonary function level on both TSP and SO_2 concentrations. Kinney et al.[3] studied the association between weekly measurements of pulmonary function and exposure to ozone (O_3) in a group of 154 school children in Kingston, Tennessee. In the majority of these children, pulmonary function level was negatively associated with O_3 concentration. The association between O_3 exposure and pulmonary function level has also been investigated in a number of summer camp studies.[4-7] In these

0-87371-573-X/95/$0.00+$.50

studies, pulmonary function was measured daily over periods of 1 to 4 weeks. Spektor et al.[4] studied a sample of 91 children who had at least seven daily pulmonary function measurements while resident at a summer camp in Fairview Lake, New Jersey. In the majority of children, O_3 was negatively associated with pulmonary function.

In all of these studies, the estimated regression coefficient of pulmonary function level on air pollution concentration varied among children. It is tempting to identify the children with the most negative slopes, or with significantly negative slopes, as a susceptible subgroup.[1] If a susceptible subgroup exists, however, one would expect children in this group to show increased susceptibility to air pollutants repeatedly when exposed to several air pollution episodes. Equivalently, the existence of a sensitive subgroup implies that the expected response to an air pollution episode varies significantly among children.

This chapter investigates the issue of heterogeneity of response. We reanalyze the data from three published studies; the Steubenville study of TSP effects,[2] the Kingston study of O_3,[3] and the Fairview Lake, New Jersey, camp study of ozone,[4] to determine whether the estimated slopes of individual regressions of pulmonary function level against air pollution concentration vary significantly more than sampling variability would predict. The basic feature of these studies is that regression analyses of pulmonary function on air pollution (or time) were performed for each child. The within-child or error variance of the regression slopes can be estimated from the deviations of the observed pulmonary function values around the individual regression lines. This error variance can be compared to the total variance of the child-specific regression slopes to see if there is evidence of response heterogeneity.

II. METHODS

A. POPULATIONS

Study populations and methods of data collection have been described elsewhere.[2-4] In Steubenville, Ohio, children participating in the Six Cities Study of Air Pollution and Health[2] were selected from four schools, by classroom, to participate in the alert study. A baseline measurement of pulmonary function was obtained in early fall or spring, before anticipated air pollution episodes. A second measurement was obtained at or immediately following an alert, and the children were then restudied on three subsequent occasions 1 week apart. The alert was triggered by a 24-h period of elevated TSP or SO_2 or by a sham alert, intended to investigate temporal variation in pulmonary function level when an alert had not occurred.[2]

In Kingston, Tennessee, children were selected from one school.[3] All but 30 of these children were participants in the Six Cities Study. Pulmonary function was measured on six occasions over a 2-month period.

The Fairview Lake study was conducted at a summer camp in which the 91 participating children were resident for 2 to 4 weeks.[4] Pulmonary function was measured daily between 11:30 A.M. and 6:30 P.M. Each child had at least seven measurement days.

All spirometric measurements in all three studies were obtained in a uniform manner by trained technicians using Collins Survey Spirometers (Warren E. Collins, Braintree, MA). In this analysis we will consider the forced vital capacity (FVC) measurements reported in all three studies and the forced expired volume in three quarters of a second ($FEV_{.75}$) reported from Steubenville and Kingston, and the similar forced expired volume in 1 s (FEV_1) reported from Fairview Lake.

B. AIR POLLUTION EXPOSURE

In the original report on the Steubenville study, each child's pulmonary function measurements were regressed on average SO_2, TSP, and temperature for the 24 h immediately preceding the pulmonary function measurements. This chapter focuses on TSP, with averaging times of 1 and 5 d preceding the pulmonary function measurement, to permit analysis of delayed and/or persistent effects. The number of averaging days is indicated as follows: TSP-1 denotes the average TSP concentration during the 1-d period preceding examination and TSP-5 denotes TSP averaged over 5 d. TSP-1 values ranged from 11 to 292 $\mu g/m^3$ and TSP-5 from 38 to 205 $\mu g/m^3$ during the Steubenville alert studies.

Regression analyses of FVC and $FEV_{.75}$ on the two measures of TSP exposure were performed for each child, taking pulmonary function growth between studies into account by using indicator variables for year of study. The regression analyses were restricted to children who participated in three or four studies, to ensure that the individual slopes were calculated from reasonable numbers of observations. (In this group, the number of observations per individual ranged from 11 to 20.) The 165 children who participated in three or four studies contributed 558 (74%) of the total of 750 observations in the original analysis.

The Kingston report disussed exposure to O_3, fine sulfates (FSO_4), and fine particulate matter (FP). This chapter focuses on exposure to O_3, defined as the maximum 1-h O_3 concentration in the 24-h period ending in the hour of lung function measurement. The Fairview Lake report presented results for O_3 exposure expressed as the mean of the hour before the pulmonary function test, plus the means of the previous 2 and 4 h.

C. STATISTICAL ANALYSIS

The variance of the individual slopes is composed of two components: random or error variation due to interoccasion variability in the pulmonary function measurements of individual children, and systematic differences in slope between individual children. The hypothesis of no heterogeneity of regression coefficients among children can be tested by calculating a variance ratio of the form

$$\frac{\sum\left[\mathrm{SSX}_i\left(\hat{b}_i-\hat{b}^*\right)^2\right]\Big/(n-1)}{\left(\sum\mathrm{SSE}_i\right)\Big/\left(\sum\mathrm{EDF}_i\right)} \tag{1}$$

where SSX_i = sum of squares of the X variable (TSP or O_3) for subject i; n = number of subjects; b_i = estimated regression coefficient for subject i;

$\hat{b}^*$ = weighted mean regression coefficient

$$=\frac{\sum\left(\mathrm{SSX}_i*\hat{b}_i\right)}{\sum\left(\mathrm{SSX}_i\right)} \tag{2}$$

SSE_i = error sum of squares for subject i; EDF_i = error degrees of freedom for subject i

The variance ratio (Equation 1) follows an F-distribution with n-1 and $\Sigma(\mathrm{EDF}_i)$ degrees of freedom when there is no between subject variability of regression slopes and will otherwise tend to be large.

Because F-tests based on approximately normally distributed observations are sensitive to observations resulting from children with highly variable pulmonary function readings, an outlier criterion was developed, and the calculations were repeated after removal of children with highly variable pulmonary function readings. Outliers were identified by calculating the pooled estimate of within-child (error) variance

$$s^2=\left(\sum\mathrm{SSE}_i\right)\Big/\left(\sum\mathrm{EDF}_i\right) \tag{3}$$

where SSE_i = sum of squares of pulmonary function values for subject i. Then, the estimated error variance for the ith child, $\mathrm{SSE}_i/\mathrm{EDF}_i$, was compared to the quantity s^{2*} ($\mathrm{chisq}_{.99}/\mathrm{EDF}_i$), where $\mathrm{chisq}_{.99}$ is the 99th percentage point of the chi-square distribution with (EDF_i) degrees of freedom. If the within-child pulmonary function variance was larger than this quantity, the child was identified as an outlier for this analysis.

Statistical analyses were performed using PC/SAS Software[8] on a Compaq Deskpro 286 personal computer.

III. RESULTS

Table 1 gives a stem-and-leaf display[9] of the distributions of the regression coefficients of FVC and $\mathrm{FEV}_{.75}$ on TSP-1 for children participating in the Steubenville

Table 1 Stem and Leaf Displays of Regression Slopes of FVC and $FEV_{.75}$ on TSP-1 for Participants in the Steubenville Alert Studies[a]

Stem	leaf	Number
FVC, mL/0.1 μg/m³		
22	1	1
20		
18		
16		
14	9	1
12		
10	3	1
8	69	2
6	125	3
4	1334668256	10
2	02233556677900222235	20
0	123344558899112333445555 88	27
– 0	766644433322111099887776544333000	33
– 2	877755533332206665544443311100	30
– 4	8766438654311110	16
– 6	81988777444331	14
– 8	241	3
– 10		
– 12	22	2
– 14		
– 16		
– 18		
– 20	9	1
– 22		
– 24	4	1
$FEV_{.75}$		
20	1	1
18		
16		
14	01	2
12	0	1
10	1	1
8	14	2
6	3684	4
4	046279	6
2	011133445567703479	18
0	223333455777888990222334455569	30
– 0	98887765555444310088877555544322111000	38
– 2	9987652211100009999986654442000	31
– 4	87520875430	11
– 6	7543521111	10
– 8	83241	5
– 10	76	2
– 12	1	1
– 14	8	
– 16		
– 18		
– 20	3	1

[a] Slopes of children with highly variable pulmonary function measurements are underlined.

Table 2 Mean slopes, averaged over studies, of FVC and $FEV_{.75}$ on TSP, full and restricted samples, Steubenville, Ohio, 1978–1980

Exposure	Pulmonary variable	Mean slope (SE) Full sample	Mean slope (SE) Restricted sample
TSP-1	FVC	– 0.099 (0.040)*	– 0.119 (0.034)†
TSP-5		– 0.192 (0.054)‡	– 0.167 (0.049)†
		n = 165	n = 154
TSP-1	$FEV_{.75}$	– 0.086 (0.040)*	– 0.104 (0.042)†
TSP-5		– 0.201 (0.059)†	– 0.174 (0.043)‡
		n = 165	n = 151

* $p < 0.05$.
† $p < 0.01$.
‡ $p < 0.001$.

Table 3 Mean slopes of pulmonary function on ozone, full and restricted samples, Kingston, Tennessee, 1981

Pulmonary variable	Mean slope (SE) Full sample	Mean slope (SE) Restricted sample
FVC	– 0.918 (0.356)*	– 0.291 (0.267)‡
	n = 154	n = 146
$FEV_{.75}$	– 0.994 (0.363)†	– 1.152 (0.300)‡
	n = 154	n = 141

* $p < 0.05$.
† $p < 0.01$.
‡ $p < 0.001$.

alert studies. The distributions show both positive and negative outlying values. All of the extremely positive and negative slopes met the criterion for large interoccasion variability. For FVC, 11 (7%) of the 165 children were found to be unacceptably variable; for $FEV_{.75}$, 14 (8%) of the 165 children were unacceptably variable. In Kingston, 8 (5%) children of a total of 154 were outliers for FVC and 13 (8%) for $FEV_{.75}$. In Fairview Lake, 7 (8%) of 91 children were outliers for FVC and 3 (3%) for FEV_1. The percentage of outliers is similar in all three studies.

Table 2 gives the mean slopes for Steubenville, both for the full sample and for the reduced sample after removing outliers. All mean slopes are negative and significantly different from zero. Removal of children with highly variable pulmonary function did not change the mean slopes much, but the standard errors of the mean slopes were generally smaller in the restricted population.

Table 3 shows the mean slopes for the Kingston data, again for the full sample and for the restricted sample. As in the Steubenville data, all mean slopes were negative and significantly different from zero. The mean slopes in the restricted sample were similar to those for the full sample, but smaller standard errors tended to make the mean slopes in the restricted sample more significant than those in the full sample, indicating increased sensitivity of the analysis after removal of children with highly variable pulmonary function.

The mean slopes of FVC and FEV_1 on O_3 for the Fairview Lake study are shown in Table 4, for the full and restricted sample. The mean slopes are all significantly

Table 4 Mean slopes of FVC and FEV_1 on ozone, full and restricted samples, Fairview Lake, New Jersey, 1984

Exposure variable	Pulmonary variable	Mean slope (SE)	
		Full sample	Restricted sample
O_3 (1 hr)[a]	FVC	– 1.291 (0.188)*	– 1.045 (0.154)*
(2 hr)		– 1.274 (0.182)*	– 1.031 (0.148)*
(4 hr)		– 1.264 (0.198)*	– 1.002 (0.160)*
		$n = 91$	$n = 84$
O_3 (1 hr)	FEV_1	– 1.401 (0.174)*	– 1.287 (0.166)*
(2 hr)		– 1.443 (0.175)*	– 1.332 (0.167)*
(4 hr)		– 1.497 (0.187)*	– 1.381 (0.177)*
		$n = 91$	$n = 88$

[a] Numbers in parentheses indicate number of averaging hours.

* $p < 0.001$.

Table 5 Variance ratios testing variability between subjects in pulmonary function response to air pollution, Steubenville, Ohio, Kingston, Tennessee, and Fairview Lake, New Jersey

Study location	F-ratios for regressions	
	FVC	$FEV_{.75/1.0}$
Steubenville		
TSP-1	1.01	0.89
TSP-5	1.12	0.98
Kingston		
O_3 (max hr)	2.62‡	1.27*
Fairview lake		
O_3 (1 hr)	1.34*	1.53†
(2 hr)	1.32*	1.48†
(4 hr)	1.33*	1.45†

* $p < 0.05$.

† $p < 0.01$.

‡ $p < 0.001$.

less than zero. As with the other two studies, removing the most variable responses reduces the standard error, but does not substantially change the mean slopes. In fact, the estimated effect of ozone in the Fairview Lake study is very close to the Kingston estimate. The standard errors are less, due in part to the larger number of observations for each child.

The comparability of the mean slopes in the full and restricted samples in each of the studies suggests that the children with highly variable pulmonary function measurements were not more responsive to air pollution exposure. If more responsive children had tended to be removed, the mean slopes would have been smaller in the restricted samples.

The variance ratio calculations (Table 5) show little evidence of heterogeneity of response between individuals in the Steubenville data; all ratios are close to 1. In Kingston, all ratios are significantly larger than 1, although only the ratio for FVC is much larger, with a value of 2.62. In Fairview Lake, all ratios are also significantly larger than 1. (The F-ratios between the studies are not directly comparable because

Table 6 Variance ratios testing variability between subjects in pulmonary function response to air pollution, after removal of outliers

Study location	F-ratios for regressions	
	FVC	$FEV_{.75/1.0}$
Steubenville		
TSP-1	0.91	0.86
TSP-5	1.07	0.89
Kingston		
O_3 (max hr)	1.45†	1.30*
Fairview lake		
O_3 (1 hr)	1.20	1.57†
(2 hr)	1.16	1.54†
(4 hr)	1.17	1.46†

* $p > 0.05$.

† $p > 0.01$.

of the different number of children and observations producing different degrees of freedom for each study.)

As noted previously, 11 of 165 Steubenville children were declared outliers in the analysis of FVC, and 14 out of 165 were declared outliers in the analysis of $FEV_{.75}$. If the variances had followed the chi-square distribution, only one or two outliers would have been expected. In the Kingston data, 8 of 154 subjects were outliers in the FVC analysis and 13 in the $FEV_{.75}$ analysis, and in the Fairview Lake data, 7 and 3 out of 91 subjects were outliers in the FVC and FEV_1 analyses, respectively. Removal of outliers decreased the significance of the F-ratios for FVC, but increased them slightly for the FEV measures (Table 6) from the Kingston and Fairview Lake studies. In the Steubenville data, all ratios are very close to 1; in the Kingston and Fairview Lake data, the ratios remain significantly larger than 1.

IV. DISCUSSION

In each of the studies considered, the distributions of the regression slopes tended to have heavy tails (Table 1). Most of the large negative and positive slopes were obtained in children with highly variable pulmonary function measurements. There was no indication that high variability of pulmonary function was associated with air pollution exposure. Children with highly variable pulmonary function had both extremely negative and extremely positive slopes. Removal of these children from the analysis resulted in distributions with less heavy tails and, consequently, smaller standard deviations (Tables 2 to 4). The percentage of children with highly variable pulmonary function measurements was larger than would have been expected, 3 to 8%, when the method would have generated only 1% if the between-occasion variablity was constant across children. The origin of this apparent high variability of pulmonary function in some of the children is not obvious, but appears to be due to factors other than air pollution episodes.

The variability of the regression slopes appeared to be largely determined by within-child variablity rather than between-child variability. In the Steubenville data, the F-ratios were close to 1 after removal of children with highly variable pulmonary function. In the two studies of ozone exposure. Kinston and Fairview Lake, there remained some evidence for heterogeneity after outlier removal, but the F-ratios were all smaller than 1.5, indicating that a large part of the observed variability was due to within-child variability and not to between-child variability. These results suggest that it is not possible to identify responders or susceptible subgroups by their position in the lower tail of an observed distribution of slopes, as has been suggested.[1]

The observed variance of response is not so much determined by real differences between subjects as by our inability to characterize individual response more exactly. One reason for this might be that the studies described here relied on central site monitoring for their assessment of exposure to air pollutants. It is well known that personal exposure may be quite different[10] from concentration measured at a central site. Variation among individuals in true air pollution exposure contributes to variability among individuals in estimated slopes. In the Steubenville and Kingston studies, which were school based, it is likely that children had very different air pollution exposures as they went about their daily activities. In the Fairview Lake study, these exposure differences were lessened by studying children in a resident camp such that the participants were always within a limited distance from the central ambient monitor.

Other studies do not suggest increased within-person variability of pulmonary function levels during air pollution episodes. Kanner et al.[11] report correlation coefficients of 0.943 (FVC) and 0.922 ($FEV_{.75}$) for pulmonary function measurements made 1 month apart in a group of 8- to 9-year-year-old children. Average correlations with baseline pulmonary function levels were 0.941 (FVC) and 0.892 ($FEV_{.75}$) in Steubenville, and 0.932 (FVC) and 0.915 ($FEV_{.75}$) in Kingston. Lower correlations would have been expected if there had been increased within-child variability due to differential response to air pollution.

McDonnell and co-workers[12] have shown that adults repeatedly exposed to ozone in a chamber had reproducible responses, but that these responses varied substantially among the individuals. Kulle and co-workers[13] also found that the dose-response curves of adults exposed to ozone in a chamber had substantially different slopes. Thus, controlled exposure studies have shown that ozone produces a reproducible pulmonary function response for individual subjects, whereas the size of the response varies among subjects.

The analysis of variance of the Kingston and Fairview Lake ozone studies suggests that the variance of the individual slopes is larger than would be expected based on the inherent variability of repeated pulmonary function measurements alone. This is consistent with the results from controlled exposure studies. For TSP concentrations, no evidence was found for clear interindividual differences in response. Thus, the results apparently differ for ozone and TSP. This may indicate a different mechanism for the effects of TSP (and associated pollutants) and ozone.

In a study of asthmatics in the Los Angeles area, Whittemore and Korn[14] found that asthma attack rates increased with oxidant and TSP concentrations after adjusting for temperature, relative humidity, day of the week, day of study, and attacks on the previous day. Interestingly, they found that the estimated TSP coefficients did not vary between individuals, but there was interindividual variability for the coefficients for ozone.

This analysis suggests that there is heterogeneity of response to ozone exposure, but not the exposure to particulate pollution episodes. Nevertheless, the component of variation in response due to random measurement error is still large, which implies that the actual range of response may be much smaller than indicated by the histograms of individual regression slopes. It is therefore not appropriate to base risk estimates for susceptible individuals on the distribution of individual response without adjusting for measurement error.

ACKNOWLEDGMENT

We are very grateful to Mort Lippmann and Dalia Spektor for allowing us to reanalyze the data collected in the Fairview Lake study. The work was supported in part by NIEHS Grants ES-0002 and ES-01108, EPA Cooperative Agreement CR 811650, and EPRI Contract RP-1001. Bert Brunekreef was supported by a grant from the Netherlands Organization for the Advancement of Pure Research (Z.W.O.), and Douglas Dockery was supported by a Mellon Foundation Faculty Development Award. This chapter has not been subjected to the Environmental Protection Agency's required peer and policy review and therefore does not necessarily reflect the views of the Agency, and no official endorsement should be inferred.

REFERENCES

1. Stebbings, J. H., Jr. and Fogelman, D. G., Identifying a susceptible subgroup: effects of the Pittsburgh air pollution episode upon school children, *Am. J. Epidemol.*, 110, 27, 1979.
2. Dockery, D. W., Ware, J. H., Ferris, B. G., Jr., Speizer, F. E., Cook, N. R., and Herman, S. M., Change in pulmonary function associated with air pollution episodes, *J. Air Pollut. Control Assoc.*, 32, 937, 1982.
3. Kinney, P. L., Speizer, F. E., Spengler, J. D., Dockery, D. W., and Ferris, B. G., Jr., Short term pulmonary function associations with ozone in Kingston, Tennessee, *Am. Rev. Respir. Dis.*, 139, 56, 1989.
4. Spektor, D. M., Lippmann, M., Lioy P. J., Thurston, G. D., Citak, K., James, D. J., Speizer, F. E., and Hayes, C., Effects of ambient ozone on respiratory function in active normal children, *Am. Rev. Respir. Dis.*, 137, 313, 1988.
5. Lippmann, M., Lioy, P. J., Leikauf, G., Green, K. B., Baxter, D., Morandi, M., Pasternak, B. S., Fife, D., and Speizer, F. E., Effects of ozone on the pulmonary function of children, in *International Symposium on the Biomedical Effects of Ozone and Related Photochemical Oxidants*, Lee, S. D., Mustafa, M. G., and Mehlman, M. A., Eds., Princeton Scientific Publishers, Princeton, NJ, 1982, 423.
6. Bock, N., Lippmann, M., Lioy, P., Munoz, A., and Speizer, F. E., The effects of ozone on the pulmonary function of children, in *Evaluation Standards*, TR-4, Evaluation of the Scientific Basis for Ozone/Oxidant Standards, Lee, S. D., Ed., Air Pollution Control Association, Houston, TX, 1984, 297.

7. Lioy, P. J., Vollmuth, T. A., and Lippmann, M., Persistence of peak flow decrements in children following ozone exposures exceeding the National Ambient Air Quality Standard, *J. Air Pollut. Control Assoc.*, 35, 1068, 1985.
8. SAS Institute Inc. SAS User's Guide: Statistics, Version 5. SAS Institute Inc., Cary, NC, 1985.
9. Emerson J. D. and Hoaglin, D. C., Stempand-leaf displays, in *Understanding Robust and Exploratory Data Analysis*, Hoaglin, D. C., Mosteller, F., and Tukey J. W., Eds., John Wiley & Sons, New York, 1983, 7.
10. Spengler, J. D., Treitmen, R. D., Tosteson, T. D., Mage, D. T., and Soczak, M. L., Personal exposures to respirable particulates and implications for air polution epidemiology, *Environ. Sci. Technol.*, 9, 700, 1985.
11. Kanner, R. E., Schenker, M. B., Munoz, A., and Speizer, F. E., Spirometry in children: methodology for obtaining optimal results for clinical and epidemiological studies, *Am. Rev. Respir. Dis.*, 127, 720, 1983.
12. McDonnell, W. F., Horstman, D. H., Abdul Salaam, S., and House, D. E., Reproductivity of individual responses to ozone exposure, *Am. Rev. Respir. Dis.*, 131, 36, 1985.
13. Kulle, T. J., Sauder, L. R., Hebel, J. R., and Chatham, M. D., Ozone response relationships in healthy non-smokers, *Am. Rev. Respir. Dis.*, 132, 36, 1985.
14. Whittemore, A. S. and Korn, E. L., Asthma and air pollution in the Los Angeles area, *Am. J. Public Health*, 70, 687, 1980.

CHAPTER 14

Health Effects from Environmental Noise Exposure

Evelyn Talbott and Shirley Jean Thompson

I. INTRODUCTION

Since the industrial revolution, rapid development of countries, particularly those with increasing industrialization and road transport, has lead to an escalation in community noise levels.[1] Noise is defined as unwanted sound. Unwanted sound to some may be interpreted as welcome sound to others, as in the case of loud rock music. In general, noise is an auditory stressor and has been shown to have several direct and indirect health effects. The only organ damage known to be directly attributable to noise exposure is from mechanical damage to the inner hair cells of the organ of corti with extremely loud noise exposure (>90 dBA). There are, however, several nonauditory effects of noise exposure that involve direct reactions to acoustical stimuli. These nonauditory physiological effects of noise include a possible increase in cardiovascular disease from elevated blood pressure, as well as physiological reactions involving the cardiovascular endocrine system. In addition to cardiovascular effects, community noise has been shown to adversely affect communication,[2] performance and behavior,[3] memory and reading acquisition,[4] sleep,[5,6] mental health,[7] and to result in annoyance reactions.[8,9]

This chapter will present a short introduction to major sources of noise, measurement of noise and sound in the environment, standards that have been set to control noise, and a description of the most recent epidemiologic studies of several possible health effects of noise from environmental sources.

II. THE PHYSICAL ASPECTS OF NOISE

Sound is produced by the vibration of air molecules and is transmitted as a longitudinal wave motion. It is, therefore, a form of mechanical energy and is

0-87371-573-X/95/$0.00+$.50

measured in energy-related units of intensity and frequency. Sound output of a source is measured in watts and the intensity of sound at a point in space is defined by the rate of energy flow per unit area measured in watts per square meter. The intensity can be shown to be proportional to the square of the amptitude of the sound wave and to the square of the frequency. The intensity of sound waves is astonishingly small. A barely audible sound has an intensity of 10^{-16} (ten million-billionths) of a watt per square centimeter. The intensities vary so greatly, their ratios are expressed by exponents. If sound "X" is 100 times (10^2) as intense as sound "Y", you say that its intensity level is 2 bels higher (a bel is a unit used to express the ratio of two powers, usually electric or acoustic powers). If "Z" has an intensity of one tenth as great as "X", the intensity level is 1 bel lower. Intensity is proportional to the mean square of the sound pressure and as the range of this variable is so wide, we usually express its values in dB (decibels).[10]

The distribution of sound energy expressed in cycles per second, or Hertz (Hz), is important because the human ear is more sensitive to sounds at some frequencies than at others. Therefore, in order to determine the magnitude of sound on a scale that is perceived by a human, it is necessary to weight the physical sound spectrum to account for the frequency response of the ear. The most universally used form of frequency weighting for sounds of all types is called A-weighting. The A-weight, which discriminates against low frequency and very high frequency sound, is a reasonably reliable and readily measured estimate of loudness.

In addition to level and frequency, duration of sound and the way it is distributed in time are important in determining noise exposures of individuals. Sounds may be continuous, varying, intermittent, or impulsive. Long-term exposure to a high level of continuous noise has been most strongly linked to health effects, but for aircraft noise, peak levels are most important.

Since the sound level scale is logarithmic, a small increase in decibels represents a large increase in sound energy. For example, if two similar noise sources operate simultaneously, the measured mean square sound pressure from the two sources will add together to give a value twice that which would result from either source alone. The resulting sound pressure level in decibels from the combined sources will be only 3 dB higher than the level produced by either source alone. (If the two sound sources produce individual levels that are different by 10 dB or more, the resulting sound pressure level approximates that of the greater source acting alone.)

III. SOUND PRESSURE AND SOUND PRESSURE LEVEL

Most sound-measuring instruments are calibrated to provide a reading of root mean square (rms) sound pressures on a logarithmic scale in decibels. The reading taken from an instrument is called a sound pressure level. The term level is used because the pressure measured is at a level above a given pressure reference. For sound measurement in air, 20 microneutons per square meter (0.00002, N/m^2) commonly serves as the reference sound pressure. This reference is an arbitrary

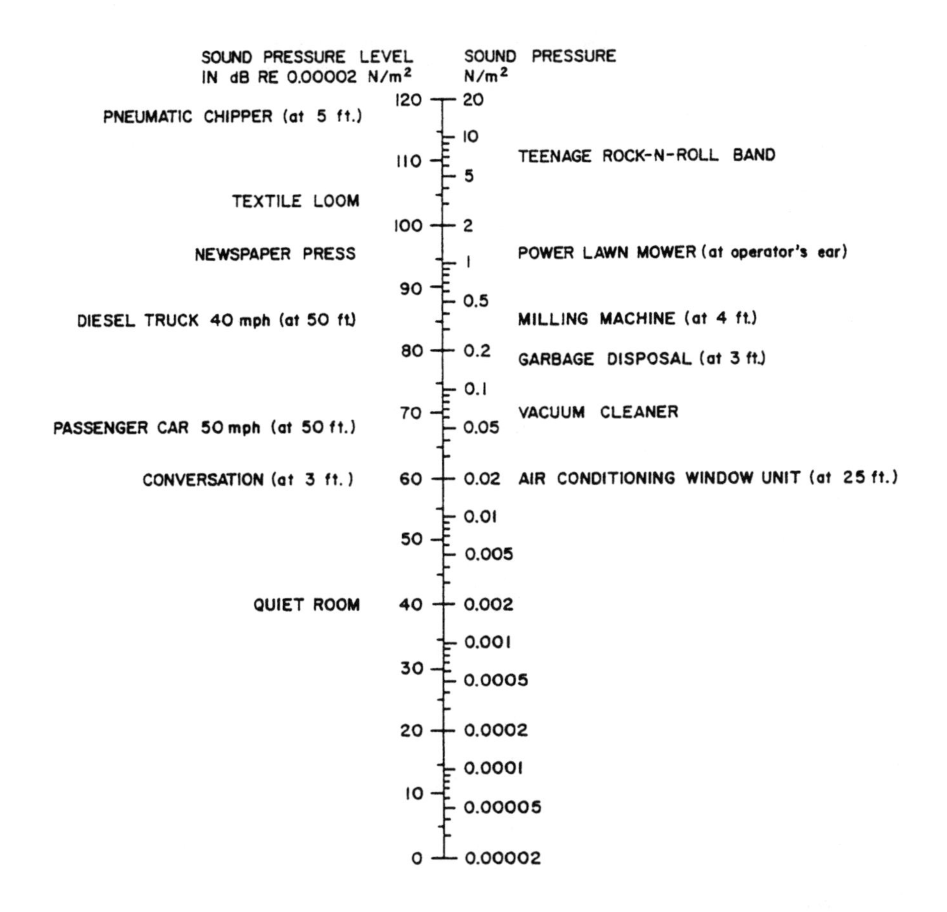

Figure 1 Relationship between A-weighted sound-pressure level in decibels (dB) and sound pressure in N/m². (From Micheal, P.L., in *Physics of Sound*, U.S. Department of Health and Human Services, Public Health Services, Washington, D.C., 1973, chap. 23.)

pressure chosen many years ago because it was thought to approximate the normal threshold of a young human hearing at 1000 Hz.[11] The mathematical form is

$$L_p = 20 \quad \log \frac{p}{p_0} dB$$

where p is measured mean square sound pressure and p_o is the reference sound pressure and the logarithm is a base 10. Thus, L_p should be written in terms of decibels referenced to a specified pressure level, for example, in air, the notation for L_p is commonly abbreviated as dB, 0.00002 N/m². Shown in Figure 1 are some relationships between the A-weighted sound pressure level and decibels (dB) and sound pressure in N/m².[11] Figure 2 presents some decibel levels for common sounds in our environment and their rank in potential harm to hearing.

SOUND LEVELS AND HUMAN RESPONSE

COMMON SOUNDS	*NOISE LEVEL (dB)*	*EFFECT*
BOOM CARS	145	Beyond threshold of pain (125 dB)
JET ENGINES (NEAR)	140	
SHOTGUN FIRING	130	
JET TAKEOFF (100-200 FT.)	130	
ROCK CONCERTS (Varies)	110-140	
OXYGEN TORCH	121	
SYMPHONY ORCHESTRA	110	
DISCOTEQUE/BOOM BOX	120	Threshold of sensation (120 dB)
THUNDERCAP (NEAR)	120	
STEREOS (OVER 100 WATTS)	110-125	
POWER SAW (CHAIN SAW)	110	Regular exposure of more than 1 min.
PNEUMATIC DRILL/JACKHAMMER	110	risks permanent hearing loss (over 100 dB)
SNOWMOBILE	105	
JET FLYOVER (1000 FEET)	103	
ELECTRIC FURNACE AREA	100	
GARBAGE TRUCK/CEMENT MIXER	100	No more than 15 min. unprotected
FARM TRACTOR	98	exposure recommended (90-100 dB)
NEWSPAPER PRESS	97	
SUBWAY, MOTORCYCLE (25 ft.)	90	Very annoying
LAWNMOWER, FOOD BLENDER	85-90	**85 - level at which hearing damage (8 hrs.)**
RECREATIONAL VEHICLES, TV	70-90	begins
DIESEL TRUCK (40 mph, 50 ft.)	84	
WASHING MACHINE	78	
DISHWASHER	75	
AVERAGE CITY TRAFFIC NOISE	80	Annoying, interferes with conversation,
GARBAGE DISPOSAL	80	constant exposure may cause damage.
VACUUM CLEANER, HAIR DRYER	70	Intrusive, intereferes with telephone use
INSIDE A CAR (LOUD ENGINE)		
GARBAGE DISPOSALS	50-60	
NORMAL CONVERSATION	50-65	
QUIET OFFICE	50-60	Comfortable (under 60 dB)
REFRIGERATOR HUMMING	40	
LIVING ROOM, BEDROOM		
WHISPER	30	Very Quiet
BROADCASTING STUDIO		
RUSTLING LEAVES	20	
NORMAL BREATHING	10	Just Audible
	0	**Threshold of normal hearing** (1000-4000 Hz)

This decibel (dB) table compares some common sounds and shows how they rank in potential harm to hearing. In many industries, workers are exposed to dangerous noise levels. This is particularly true in the construction, lumber, mining, steel and textile industries.

Since the sensitivity of the ear to sound is not the same for all frequencies, weighting or attenuating filters are included in the sound level meter's circuits to simulate the ears' response. A noise level meter gives an instantaneous measurement of the noise present, but cannot measure the *duration* of the exposure. To measure the amount of noise a person is exposed to over a period of time, a "dosimeter" or an integrated sound level meter must be used. Sources for above include the *American Medical Association* and the *Canadian Hearing Society of Ontario.*

Decibel table developed by the National Institute on Deafness and Other Communication Disorders, National Institutes of Health, Bethesda, Maryland 20892. January, 1990.

Figure 2 Sound levels and human response.

IV. SOURCES OF NOISE

There are approximately 8 million people in the U.S. who are exposed to occupational noise levels of 80 dBA or greater, 4,685,000 exposed to levels of 85 dBA or greater, and 2,707,000 exposed to levels of 90 dBA or greater.[12] Clearly, occupational noise exposure is the most significant risk factor for noise-induced hearing loss in our society. However, the extra occupational or nonoccupational sources of noise exposure include community noise exposure from installation and construction sites, which often add to annoyance and communication difficulties; road traffic, particularly near schools and other public buildings; and rail traffic. In addition, aircraft operations have caused severe community noise problems over the past 20 to 30 years, particularly in densely populated areas near military and nonmilitary commercial airports. Sonic booms generated from low-flying military aircraft, as well as construction and public works from building construction, cause considerable noise emissions. There is domestic noise related to television sets, hobbies, stereo systems, noise from leisure activities such as motorboats, water skiing, car stereos, and outdoor rock concerts. In all, levels of noise in our society are likely to go up, not down, as the density of population increases.

There are several sensitive subsets of the population, including young children, older adults, and individuals with chronic conditions, which may be more adversely affected by the hazards of noise.[1] Because of this, a series of standards have been promulgated to protect individuals from noise.

A. PRESENT NOISE STANDARDS

Presently in the U.S., several pieces of legislation address noise standards in the workplace. They include the Occupational, Safety and Health Act of 1970, the Noise Control Act of 1972, the Mine Safety Act of 1978, and the Noise Amendment of 1983.[13,14] These acts require certain agencies to regulate noise. The present effective noise standard, 29 CFR 1910.95, states a maximum time-weighted exposure level of 90 dB(A) with a 5-dB trading relation, that is, a doubling factor between exposure time and intensity; this means that 4 h of exposure to 95 dBA is considered to be equally hazardous as 90 dBA for 8 h; or 2 h at 100 dB or 1 h at 105 dB; 30 min at 110 dB or 15 min at 115 dB. A doubling factor of 3 dB implies the same total energy whether the exposure consists of a short time in high-level noise or a long time in low-level noise (3-dB increase in sound is the same as a doubling of the energy). A 5-dB doubling means that it is assumed that a larger total energy can be tolerated if it is presented during a short time or during brief periods of the working day. The results of recent research indicate that this is not the case for the doubling factor of 3-dB, which may be adequate for relatively low noise levels. A smaller doubling factor may more correctly reflect the hazard of noise at a high level. The U.S. Standard also states no exposure is allowed to continuous noise above 115 dBA or impulse noise above 140 dBA. Many countries use the International Standards Organization (I.S.O.) Recommendation as a basis for noise legislation. The I.S.O. interprets the amount of risk for requiring a hearing loss of 25 dB average at 0.5, 1, and 2 kHz as a result of exposure to certain noise levels. Many countries have chosen the 85-dBA level for 8 hr exposure and assume the 3-dB doubling factor.[15]

In 1972, the National Institute of Occupational Safety and Health (NIOSH) published the recommendation that the standard for occupational exposure to noise should be 85 dBA for an 8-hr TWA instead of 90 dBA. In 1983, the Occupational Safety and Health Administration (OSHA) announced a hearing conservation amendment to the 1972 noise standard. This amendment defined effective hearing conservation and required that a hearing conservation program be administered if the work noise level goes to 85 dBA or greater. This regulation states that a hearing conservation program is required to include an assessment for hearing protection.[15]

Between 1973 and 1978, the percentage of the population in the U.S. exposed to unacceptable road traffic noise almost doubled.[16,17] About 50% of the population is exposed to unacceptable traffic noise of greater than 65 dBA. In larger cities, more than half are exposed to more than this level. Given this information, it is understandable that researchers have focused on the question of what are the medically relevant effects of noise from community noise exposure.

Community annoyance resulting from noise-induced activity interference was one of the most important considerations in the Environmental Protection Agency's (EPA's) identification of an outdoor DNL (day-night average sound level) of 55 dB as the "safe" level of environmental noise.[18,19] Some years later, a Federal Inter-Agency Committee on Urban Noise (FICUN) developed guidelines for considering noise in land-use planning and control.[20]

In its noise zone classification table, "minimal" exposures to noise were defined as DNLs below 55 dB, and between DNLs of 55 and 65 dB, the exposures were labeled "moderate". However, all of these exposures were considered "acceptable" according to land-use planning standards specified by the Department of Housing and Urban Development (HUD).[32]

The Department of Transportation's Federal Aviation Administration (FAA) has adopted a DNL of 65 as the point above which residential land-use becomes "normally unacceptable". Below this level, the FAA does not require airport authorities to draw noise contours or discuss the impact of airport noise on the surrounding communities for purposes of compatibility planning or to receive grants under the Part 150 program. Thus, public policy decisions, at least on the federal level, have not considered the annoyance of individuals living in the DNL 55- to 65-dB range.[20]

V. HEALTH EFFECTS

A. HEARING LOSS

At exposure levels below 80 dB, an increased risk of hearing loss caused by occupational noise exposure has not been found. Based on average hearing thresholds at 1000, 2000, and 3000 Hz, OSHA estimates that about 17% of production workers or 1.6 million people have at least mild hearing loss resulting from occupational noise exposure, 1 million or 11% have moderate hearing impairment, and

Table 1 Risk Factors for Noise-Related Hearing Loss

Primary Factors

1. Occupational noise exposure (85 dBA)
2. Military history
3. Noisy hobbies
    a. Musical instruments (percussion, brass, etc.)
    b. Outdoor target shooting/hunting
    c. Gardening with gas mower, etc.
    d. Work with cars/engines
    e. Woodworking/power tools
4. Medication usage (may initiate as well as potentiate hearing loss)
    a. Kanamycin, cisplatin
    b. Aspirin
    c. Diuretics
5. Age (may initiate or potentiate hearing loss)
6. Residential factors (highly urban environment, outdoor activities)

Secondary Factors (Lifestyle/Individual Factors)

1. Anatomic and genetic factors (tough vs. tender ear hypothesis)
    a. History of hearing loss in family
    b. Other possible individual differences in inner and outer ear anatomy and physiology
2. Diabetes (peripheral neuropathy and increased atherosclerosis of small vessels)
3. Elevated cholesterol (increased atherosclerosis may affect small vessels)
4. Cigarette smoking (increased atherosclerosis may affect small vessels)
5. Elevated blood pressure

From Talbott, E., Presented at NIH Consensus Development Conference on "Noise and Hearing Loss", National Institutes of Health, Bethesda, Maryland, January 22-24, 1990.

473,000 or 5% have moderate to severe hearing impairment. Other estimates indicate one fourth of persons 55 years of age or older who have been exposed over their working lifetime to an average of 90 dBA have developed a moderate hearing impairment caused by occupational noise exposure.[13,14] In addition to vocation, other factors such as military history, noisy hobbies, certain factors such as medication usage, as well as presbycusis or the effect of age on hearing, are additional risk factors for noise-related hearing loss (see Table 1). In addition to these primary risk factors, there are certain individual and lifestyle factors that may increase an individual's risk of noise-related hearing loss. These include anatomic and genetic factors, diabetes, elevated cholesterol, cigarette smoking, as well as high blood pressure, which may affect the small arteries and vessels that provide circulation to the ear. According to the National Institutes of Health's (NIH) Consensus Development Conference on Noise and Hearing Loss, the cohort of individuals that are presently in their teens and early twenties have been exposed to the heightened effects of electronic media, i.e., walkmans, rock concerts, etc., and are at increased risk of high-frequency hearing loss.[22]

B. CARDIOVASCULAR EFFECTS

Elevated blood pressure, because it is such an important predictor of heart disease, has been the major nonauditory health effect of noise studied, first in occupational and, more recently, in community settings. Overall, this research indicates that long-term exposure to high-intensity noise at 85 dBA and above, especially when the ears have been unprotected, may lead to sustained blood pressure elevations or hypertension.[21,23] This association is evident when there is a long duration of noise exposure even after adjusting for other major cardiovascular risk factors. An example of such research is a study covering the range of noise from 75 to 104 dBA in a textile mill, which demonstrated a dose-response relationship between level of noise and prevalence of hypertension among 1101 women who had worked in a specific shop with unprotected ears for their entire working lives.[24] The odds of hypertension increased by 1.2 for each 5-dBA increase in noise (odds ratio = 1.8 at 95 dBA), after adjusting for age, working years, salt intake, and family history of hypertension. Noise was third in order of importance as a determinant of hypertension after family history and high salt intake. However, the effect of noise on blood pressure has been weak in case-control and prospective studies when confounding variables have been controlled and hearing protectors considered.[21] These findings suggest that exposure to continuous noise at levels that protect hearing may protect against nonauditory effects of industrial noise, but this hypothesis needs to be directly tested in prospective studies.

Community (as opposed to industrial) noise has received increasing attention as a potential stressor on the cardiovascular system as larger percentages of the population are exposed to aircraft and traffic noise during both day and night hours. Although this research has been less persuasive than the industrial noise studies, findings are suggestive of weak to moderate effects of community noise on blood pressure.[25-27] A major research effort has been in progress in Caerphilly (Wales) and Speedwell (England) since the mid-1980s.[25,28,29] In these communities, 2512 and 2348 men, respectively, aged 45 to 59, were assessed on noise exposure and an array of cardiovascular risk factors and followed prospectively to determine the long-term effects of road traffic noise. Subjects were grouped in 5-dB categories of traffic noise levels, ranging from 51 to 70 dBA. When the data were pooled and the lower noise exposed group (noise 60 dBA) compared to the highest exposure group (noise 66 to 70 dBA), the relative risk for ischemic heart disease was 1.1 and not statistically significant. In Caerphilly, this association was more pronounced in men who were also exposed to noise at work. These preliminary findings from this ongoing prospective study are consistent with those from population-based case-control studies in Berlin and from the Amsterdam Air Traffic Noise Study. In the Berlin study, noise levels were somewhat higher, ranging from 60 dBA to 80 dBA. The relative risk for myocardial infarction after adjustment for confounders was 1.2. These data reflect very similar risk estimates compared to both the Bonn, Germany, Traffic Noise Study and the Amsterdam Aircraft Noise Study reported earlier in the literature. Overall studies tend to indicate that, above levels of 65 dBA, community noise can affect both the incidence and prevalence of ischemic heart disease in exposed populations. Although community noise studies have traditionally considered only noise from a single source such as aircraft, it is becoming increasingly apparent that the total amount of daily noise exposure from multiple sources and the opportunity for quiet periods during which physiological systems can recover may greatly influence the individual's long-term response to the noise stimuli. At present it is not clear whether disturbances to sleep from nocturnal noise have lasting effects on the cardiovascular system. It is well documented that exposure to aircraft noise at night results in transitory increases in blood pressure, heart rate, vasoconstriction, endocrine reactions, and change in respiration during sleep.[5,6] Thus, from a public health standpoint, in large urban populations the small observed health effect may be increasingly important as the number of noise sources to which an individual is exposed continues to multiply.

In addition to the characteristics of noise itself, individual differences appear to be responsible for determining a person's response to noise, and these in turn, may shape the nature of the observed nonauditory health effect of noise. Perceived control over noise, the necessity and importance of the source of noise and its predictability, coping strategies, genetic traits, and general noise sensitivity are among the factors that may interact to modify the physiological response of an individual to high noise. Very large samples are required for more definitive statistical proof of an effect when these factors are taken into account.

C. ANNOYANCE AND GENERAL WELL-BEING

Exposure to environmental noise may also result in annoyance. Annoyance can be defined as a subjective feeling of displeasure or dissatisfaction that results from an agent known to or presumed to affect health and well-being. In terms of numbers of individuals affected, annoyance is probably the more widespread reaction to noise. As summarized by Fidell et al.,[8,9] most studies show an increase in annoyance with noise level. Aircraft noise annoyance, in average vs. maximum noise levels, were studied by Björkman et al.[30] in five areas located at various distances from the runway in Gothenburg, Sweden, to determine the extent of annoyance in residential areas that surrounded the airports. The noise exposure was assessed as the time-weighted equivalent level or as MNL (maximum noise level) and number of overflying aircraft. The results supported the conclusion that the annoyance reaction is better related to the number of aircraft and maximum noise level than to energy equivalent levels for noise exposure.

Both noise annoyance and noise sensitivity have been associated with psychiatric disorder.[7] Noise sensitivity scales include annoyance as one component. Recently, Stansfeld et al.[31] addressed the question of road traffic noise, noise sensitivity, and psychiatric disorders. Earlier cross-sectional studies have related noise exposure to psychological symptoms, but have been criticized by the choice of low-noise control groups as well as response bias related to knowledge of the survey purpose. Psychiatric disorder was measured in a Caerphilly sample by using the general health questionnaire and noise sensitivity by the 10-item Weinstein scale. In the cross-sectional study, there was no association between noise exposure and psychiatric disorder, except among those in intermediate and low noise sensitivity where rates of psychiatric disorder increased with increasing noise level. In the highest noise sensitivity group, there were high levels of psychiatric disorder regardless of noise. This might be explained by noise sensitivity partially measuring the tendency to over-report the symptoms. Initial prospective analysis has found no association between traffic noise level at baseline and later psychiatric disorder but has found a small significant association between noise sensitivity and the risk of future psychiatric disorder. The authors conclude that it appears that noise sensitivity may be an indicator of future vulnerability to psychiatric disorder, but does not seem to confer psychological vulnerability to the effects of traffic noise in this sample.

In conclusion, the major health effects of occupational as well as community noise exposure are noise-induced hearing loss, increased blood pressure, annoyance, and possible psychiatric disorders.

REFERENCES

1. Berglund, B., Ed., W.H.O. Criteria Document on Community Noise (External Review Draft), June 28, 1993, World Health Organization, Geneva, Switzerland.
2. Lazarus, H., Prediction of verbal communication in noise — a review, *Appl. Acoust.*, 19, 439, 1986.
3. Smith, A.W., A review of the effects of noise on human performance, *Scand. J. Psychol.*, 30, 185, 1989.
4. Green, K.B., Pastermack, B., and Shore, R., Effects of aircraft noise on reading ability of school-age children, *Arch. Environ. Health*, 37, 24, 1982.
5. Griefahn, B. and Gros, E., Noise and sleep at home, a field study on primary and after effects, *J. Sound Vib.*, 105, 373, 1986.
6. Ohrstrom, E., Sleep disturbance, psycho-social and medical symptoms — a pilot survey among persons exposed to high levels of road traffic noise, *J. Sound Vib.*, 133, 117, 1989.
7. Stansfeld, S.A., Noise. Noise sensitivity and psychiatric disorder: epidemiological and psychophysiological studies, *Psychol. Med., Monogr. Suppl.*, 22, 44, 1992.
8. Fidell, S., Barber, D., and Schultz, T.J., Updating a dosage-effect relationship for the prevalence of annoyance due to general transportation noise, *J. Acoust. Soc. Am.*, 89, 221, 1991.
9. Fidell, S., Schultz, T.J., and Green, D.M., A theoretical interpretation of the prevalence rate of noise-induced annoyance in residential populations, *J. Acoust. Soc. Am.*, 84, 2109, 1988.
10. Blackwood, O., Kelly, W.C., and Bell, R., *General Physics*, John Wiley & Sons, New York, 1965, chap. 22.
11. Michael, P.L., The industrial environment — its evaluation and control, in *Physics of Sound*, U.S. Department of Health and Human Services, Public Health Services, C.D.C., N.I.O.S.H., Superintendent of Doc., U.S. Government Printing Office, Washington, D.C., 1973, chap. 23.
12. Franks, J.R., Number of workers exposed to occupational noise, *Semin. Hearing*, 9(4), 287, 1988.
13. National Institute of Occupational Safety and Health (NIOSH), Criteria for a recommended standard — occupational exposure to noise, NIOSH Pub. No. HSM 73-11001, Department of Health, Education, and Welfare, Health Services and Mental Health Administration, Washington, D.C., 1972.
14. Occupational Safety and Health Administration, Final regulatory analysis of the hearing conservation amendment, GPO No. 723-860/752 1-3, Department of Labor, Washington, D.C., 1981.
15. Møller, Aage R., Noise as a health hazard, in *Effects of Physical Environment in Public Health Preventive Medicine*, 11th ed., Last, J.M., Ed., Appleton Century-Crofts, New York, 1983, chap. 16.

16. Environmental Protection Agency (EPA) Noise in America: The Extent of the Noise Problem, EPA/9-81-101, U.S. Environmental Protection Agency, Washington, D.C., 1981.
17. Environmental Protection Agency (EPA), Protective Noise Levels: Condensed Version of EPA Levels Document, EPA 550/9-79-100, U.S. Environmental Protection Agency, Washington, D.C., 1978.
18. Environmental Protection Agency (EPA), Information on Levels of Environmental Noise Requisite to Protect Public Health and Welfare with an Adequate Margin of Safety, EPA 550/9-74-004, U.S. Environmental Protection Agency, Washington, D.C., March 1974.
19. Environmental Protection Agency (EPA), Protective Noise Levels, Condensed Version of EPA Levels Document, EPA 550/9 79-100, U.S. Environmental Protection Agency, Washington, D.C., 1978.
20. Federal Interagency Committee on Urban Noise, Guidelines for Considering Noise in Land Use Planning and Control, 1981-338-006/8071, U.S. Government Printing Office, Washington, D.C., June 1980.
21. Thompson, S.J. and Fidell, S., A review of epidemiologic studies of the effects of hearing protection devices on cardiovascular response to noise exposure, Proceedings of the Hearing Conservation Conference, University of Kentucky, Lexington, Kentucky, 40506-0046, April 1992, pp. 111-118.
22. Clark, W.W., Noise exposure from leisure activities: a review, *J. Acoust. Soc. Am.*, 90, 175, 1991.
23. Talbott, E.O., Kuller, L.H., Matthews, K., Helmkamp, J., and Cottington, E., Occupational noise exposure and the epidemiology of high blood pressure, *Am. J. Epidemiol.*, 121, 501, 1985.
24. Zhao, Y., Zhang, S., Selvin, S., and Spear, R.C., A dose-response relation for noise induced hypertension, *Br. J. Ind. Med.*, 48, 179, 1991.
25. Babisch, W., Elwood, P., and Ising, H., Road traffic noise and heart disease risk: results of epidemiological studies in Caerphilly, Speedwell and Berlin, in Noise as a Public Health Problem (Proceedings of 6th International Congress), July 5-9, 1993. 1:23, Nice, France.
26. Schwarze, S. and Thompson, S.J., Research on non-auditory physiological effects of noise since 1988: review and perspectives, Noise as a Public Health Problem (Proceedings of the 6th International Congress), July 5-9, 1993, France.
27. Berglund, B., Lindvall, T., and Nordin, S., Adverse effects of aircraft noise, *Environ. Int.*, 16, 315, 1990.
28. Babisch, W., Elwood, P.C., Ising, H., and Kruppa, B., Traffic noise as a risk factor for myocardial infarction, in *Noise and Disease,* edited by H. Ising and B. Kruppa, Gustov Fisher Verlad, Stuttgart/NY, 1993, 158-178.
29. Babisch, W., Ising, H., Gallacher, J.E.J., and Elwood, P.C., Traffic noise and cardiovascular risk. The Caerphilly Study, First Phase, Outdoor Noise Levels and Risk Factors, *Arch. Environ. Health*, 43(6), 407, 1988.
30. Björkman, M., Åhrlin, U., and Rylander, R., Aircraft noise annoyance and average versus maximum noise levels, *Arch. Environ. Health*, 47(5), 326, 1992.
31. Stansfeld, S., Gallacher, J., Babisch, W., and Elwood, P., Road Traffic Noise, Noise Sensitivity, and Psychiatric Disorder, Preliminary Prospective Findings from Caerphilly Study in Noise as a Public Health Problem (Proceedings 6th International Congress), Vol. 1., page 24, Nice, France, July 5-9, 1993.
32. Suter, A.H., Noise and Its Effects. Administrative Conference of the United States, November, 1991.

Index

A

B

C

D

E

F

G

H

I

J

L

M

N

O

P

Q

R

S

T

U

V

W

X

Y